# Genetic Engineering of Plants

*Agricultural Research Opportunities and Policy Concerns*

Board on Agriculture
National Research Council

NATIONAL ACADEMY PRESS
Washington, D.C.   1984

NATIONAL ACADEMY PRESS   2101 CONSTITUTION AVENUE, NW   WASHINGTON, DC 20418

The convocation from which this publication was developed was approved by the Governing Board of the National Research Council, whose members are drawn from the councils of the National Academy of Sciences, the National Academy of Engineering, and the Institute of Medicine.

The National Research Council was established by the National Academy of Sciences in 1916 to associate the broad community of science and technology with the Academy's purposes of furthering knowledge and of advising the federal government. The Council operates in accordance with general policies determined by the Academy under the authority of its congressional charter of 1863, which establishes the Academy as a private, nonprofit, self-governing membership corporation. The Council has become the principal operating agency of both the National Academy of Sciences and the National Academy of Engineering in the conduct of their services to the government, the public, and the scientific and engineering communities. It is administered jointly by both Academies and the Institute of Medicine. The National Academy of Engineering and the Institute of Medicine were established in 1964 and 1970, respectively, under the charter of the National Academy of Sciences.

83-063138

Library of Congress Catalog Card Number 83-63138

International Standard Book Number 0-309-03434-5

Printed in the United States of America

*Cover Photograph:* Rice farming scene in northern California. © 1982 by William Garnett, reproduced with permission of the photographer.

# Convocation on Genetic Engineering of Plants: Speakers and Panelists

*Introductory Remarks*

WALTER A. ROSENBLITH, Foreign Secretary, National Academy of Sciences
GEORGE A. KEYWORTH II, Science Adviser to the President
GEORGE E. BROWN, JR., Chairman, House Subcommittee on Department Operations, Research and Foreign Agriculture, U.S. Congress

*Why Agricultural Research?*

ORVILLE BENTLEY, Assistant Secretary for Science and Education, U.S. Department of Agriculture *(National Perspective)*
W. DAVID HOPPER, Vice President, South Asia Division, World Bank *(International Perspective)*
RALPH W. F. HARDY, Director of Life Sciences Research, E. I. du Pont de Nemours & Co., Inc. *(Industrial Perspective)*
LOWELL N. LEWIS, Assistant Vice President for Agriculture and Director, California Agricultural Experiment Station System *(University Perspective)*

*A Science Overview of Genetic Engineering*

LAWRENCE BOGORAD, Biology Department, Harvard University

*Research Opportunities*

ROBERT GOLDBERG, Biology Department, University of California at Los Angeles *(Gene Structure, Function and Genetic Transmission)*
WILLIAM SCOWCROFT, Plant Industries Division, CSIRO, Canberra, Australia *(Somaclonal Variation, Cell Selection and Genotype Improvement)*
CHARLES J. ARNTZEN, Director, Plant Research Laboratory, Michigan State University *(Introducing Herbicide Resistance into Crops via Novel Genetics)*
MILTON N. SCHROTH, Department of Plant Pathology, University of California at Berkeley *(Organisms in the Rhizosphere)*

*Perspectives on Future Research (Panel Discussion)*

RONALD L. PHILLIPS, Department of Agronomy and Plant Genetics, University of Minnesota

STEPHEN P. BAENZIGER, Agricultural Research Service, Beltsville

WINSTON BRILL, University of Wisconsin and Cetus-Madison, Inc.

GLENN B. COLLINS, Agronomy Department, University of Kentucky

ROBERT M. GOODMAN, Calgene, Inc.

KEITH A. WALKER, Plant Genetics

*An Overview of Policy Issues*

VERNON W. RUTTAN, Department of Agricultural and Applied Economics, University of Minnesota *(Agricultural Productivity Implications of Biotechnology Research)*

ANTHONY J. FARAS, University of Minnesota and Co-chairman, Molecular Genetics, Inc. *(Economic Considerations in Founding a Biotechnology Company)*

RENE TEGTMEYER, Assistant U.S. Commissioner of Patents, Patent and Trademark Office, U.S. Department of Commerce *(Patents)*

RAYMOND THORNTON, President, Arkansas State University *(Safety Regulations for Recombinant DNA)*

*Education and Manpower (Panel Discussion)*

CHARLES HESS, Dean, College of Agriculture and Environmental Science, University of California at Davis

MARY E. CLUTTER, Division of Physiology, Cellular and Molecular Biology, National Science Foundation

PHILIP FILNER, ARCO Plant Cell Research Institute

KENNETH J. FREY, Agronomy Department, Iowa State University

*The Roles of Universities, Industry, and Government: Opportunities for Cooperation and Collaboration*

GILBERT S. OMENN, Dean, School of Public Health and Community Medicine, University of Washington *(Overview of Types of Agreements: University-Industry-Government)*

REUVEN M. SACHER, Director, Biological Research, Monsanto Agricultural Products Co. *(Industry's Roles)*

THEODORE HULLAR, Director of Research, Cornell University *(Government's Roles)*

RICHARD S. CALDECOTT, Dean, College of Biological Sciences, University of Minnesota *(University's Roles)*

*Closing Remarks*

WILLIAM L. BROWN, Chairman, Board on Agriculture, National Research Council

# Acknowledgments

We wish to express particular appreciation to Leslie Roberts who wrote this summary report. Also, we acknowledge the staff of the Council for Research Planning in Biological Sciences, Inc., Claire Wilson and Carol Von Dohlen, and those with the Board on Agriculture of the National Research Council, Beulah Bresler, Jill Curry, and Jim Tavares, who provided valuable assistance in organizing the meeting and in producing this report.

Finally, we sincerely appreciate the tireless efforts of Alexander Hollaender in helping to arrange the convocation and the contributions made by the speakers, panelists, and other participants without whom the convocation and report would not have been possible.

The convocation and study was supported by grants to the National Research Council's Board on Agriculture from the NAS Academy-Industry Program; National Science Foundation; U.S. Department of Agriculture; and U.S. Department of Energy; and by grants to the Council for Research Planning in Biological Sciences from Allied Corporation; ARCO Plant Cell Research Institute; Calgene, Inc.; DeKalb-Pfizer Genetics; E. I. du Pont de Nemours & Co., Inc.; Monsanto Company; North American Plant Breeders; Pfizer, Inc.; Pioneer Hi-Bred International, Inc.; Plant Genetics; The Standard Oil Company of Ohio; University of California, Davis.

# Board on Agriculture

# Preface

U.S. agriculture has flourished in this century. In the past 50 years, agricultural productivity has more than doubled, while the amount of land under cultivation has declined. This phenomenal growth has been fueled by advances in agricultural science—by farm mechanization and the development of agricultural chemicals and improved plant varieties.

Yet productivity gains may be harder to achieve in the next 50 years. In the United States, there is little prime farmland left to develop. Even with increased use of fertilizer and pesticides, farmers may not be able to squeeze more productivity from existing varieties. For some crops, yields are beginning to level off. Some of the richest croplands are becoming salinized; topsoil is being lost. And farmers are confronting rising costs for fuel, fertilizer, and water.

In much of the Third World, hunger is pervasive. In many developing nations, farmers still struggle to extract an adequate food supply from marginal soils—and world food production must somehow double in the next 40 years to meet the expected rise in demand.

To sustain agricultural productivity, farmers worldwide will need new technologies, especially new plant varieties adapted to harsh conditions. Throughout the agricultural community, there is an increasing awareness that genetic engineering may provide some of the solutions. Recombinant DNA and the other genetic engineering techniques, developed in animals in the early 1970s, have only recently been applied to plant research. Yet already, they have yielded a wealth of information about the basic structure and function of plant cells—information that can be channeled into more effective breeding strategies. And eventually, it may be possible to manipulate plant genes in the laboratory to improve plants and perhaps create new ones.

At this stage, research effort is still largely centered in basic research laboratories—though excitement has spread throughout the agricultural research community. In just a few years, the scientific progress of a few plant scientists has been remarkable. In many ways, their work constitutes a revolution—one that foretells of changes that the agricultural enterprise may undergo as it absorbs the new technology.

On May 23-24, 1983, plant scientists and science policymakers from government, private companies, and universities met at the National Academy of Sciences for a convocation on the genetic engineering of plants. Organized with the aid and encouragement of Dr. Alexander Hollaender of The Council for Research Planning in Biological Sciences, Inc., the convocation was cosponsored by Dr. Hollaender's organization and the Board on Agriculture of the National Research Council.

During the convocation, the researchers described some of the ways genetic engineering may be used to address agricultural problems. Policymakers delineated and debated the changes in research funding and training necessary to realize this potential. Various speakers urged new collaborative efforts among basic scientists and plant breeders. And all spoke of a new era in agricultural research.

This book summarizes those discussions. It is intended to serve as an introduction and guide for those who wish to follow the development of this promising new technology.

WILLIAM L. BROWN
*Chairman*, Board on Agriculture

# Contents

# Genetic Engineering of Plants

*Agricultural Research Opportunities and Policy Concerns*

# Introduction

Farmers have been cultivating plants for millenia. Over the years, increasingly sophisticated tools have been applied to crop improvement. Modern plant breeding techniques, for instance, have been used to create new varieties adapted to specific environments or needs—such as crops that are easier to harvest or that are resistant to disease. These breeding practices have been augmented by chemical technology. Pesticides are now widely used to protect crops from insect infestations. Herbicides have largely replaced mechanical cultivation as the method for controlling weeds. And fertilizer is routinely used to replenish the nutrients lost from the soil. These breeding strategies, agricultural chemicals, and improved cropping practices have boosted agricultural productivity in the United States to its current high.

Despite these successes, crops are still lost to pests, diseases, and climatic extremes. Fertilizer and other chemicals are now consuming an increasing share of the farm dollar. At the same time, there is growing concern about the effect of these chemicals on the environment. Agricultural problems are more pronounced in the Third World, where the population is rising steadily. Without improvements in agricultural technologies, demand for food is expected to outrun supply early in the next century.

Molecular biology and genetic engineering offer new tools to meet these and other agricultural needs. Molecular biologists are learning how to transfer foreign genes into plant cells. They are doing on a molecular level what plant breeders have been doing with whole plants for centuries: combining genes in new ways to create improved crops. Working with single genes, rather than whole plants, offers several advantages. One advantage is specificity. Classical breeding introduces genes that complicate the process of crop improvement. In a sexual cross, the entire

1

genomes of two plants are combined even though the breeder may be trying to transfer a trait controlled by a single gene. It takes repeated back crosses to eliminate the extraneous genes and thus many years to create an improved variety. Using molecular techniques, a gene can be snipped from one plant and spliced into another in a single experiment. Perhaps more important, genetic engineering opens up a new source

of genetic variability that can be used in crop improvement. Breeders can work only with plants that are cross-fertile. By contrast, genetic engineering offers the promise of selecting valuable traits from any organism. For instance, research is already under way on the transfer of the genes for nitrogen fixation from bacteria to plants. Another approach might be the transfer of genes for herbicide resistance from weeds to crop plants.

If genetic engineering techniques can be mastered, they could be used in the design of plants that are hardier, higher yielding, more nutritious, or less expensive to produce—such as plants that require fewer pesticides, fungicides, or fertilizers. Other possibilities include plants that can thrive in marginal conditions, on soils that are too salty, too acidic, too wet, or too dry.

The genetic engineering of plants, however, is still in its infancy. Fundamental questions remain about the feasibility of some of these techniques. Last year, researchers demonstrated for the first time that a foreign gene can be successfully inserted into a plant and made to function. Yet extensive research will be necessary before these techniques can be used in practical crop improvement schemes. Molecular biologists must identify agriculturally important genes from the 5 million or so contained in most plants. They are searching for vectors to carry foreign genes into a plant cell. And they must develop reliable methods for regenerating plants from single cells in culture. Such experiments are just beginning. In addition, little is known about how plants will respond to the introduction of foreign genes—if, for instance, yield or vigor will suffer.

Progress in plant genetic engineering has been hampered by the limited knowledge of plant biology. The successful application of genetic engineering to plants will require fundamental breakthroughs in the understanding of gene expression and regulation, as well as increased knowledge of plant physiology, biochemistry, development, and metabolism. Yet relatively few scientists are trained in either plant molecular or cellular biology, and few of these scientists have any experience in addressing agricultural questions.

It is too early to assess with accuracy either the potential or limitations of genetic engineering for crop improvement. At this stage, gene transfer is not expected to have a significant effect on agricultural production practices until the late 1990s. Other, simpler techniques, based on the ability to culture and regenerate plant cells, are already proving a shortcut in the selection and breeding of some crops. Yet in the near term, the biggest contribution of gene transfer and other new technologies will be to fundamental knowledge. The ability to identify and isolate single genes is a valuable tool in the study of gene structure and function. This knowledge can then be used to devise more effective strategies for crop improvement through classical breeding and, eventually, through genetic engineering.

In May 1983 the Board on Agriculture of the National Research Council held a convocation to discuss the potential contribution of genetic engineering to agriculture. Speakers from numerous disciplines discussed the research opportunities presented by the new genetic technologies, as well as their implications for funding and training in the plant sciences. The following report is based on that discussion.

# Crop Improvement

Crop improvement, the engineering of plants for the benefit of humanity, is as old as agriculture itself. Some 10,000 years ago, primitive people made the transition from hunting and foraging to cultivating crops. With that switch began the continuous process of improving the plants on which we depend for food, fiber, and feed.

Throughout the milennia, two techniques have been used to improve crops, according to Lawrence Bogorad, a plant molecular biologist at Harvard University. The first is selection, which draws on the genetic variation inherent in plants. The earliest farmers selected plants having advantageous traits, such as those that bore the largest fruit or were the easiest to harvest. Perhaps through some rudimentary awareness that traits were passed from one generation to the next, the choicest plants and seeds were used to establish the next year's crop. Natural selection, which determines the survival of species, was now augmented by artificial selection. By selecting and isolating choice plants for cultivation, the early farmers were in essence influencing which plants would cross-pollinate. Through selection and isolation, they were narrowing, yet controlling, the available gene pool for each crop.

Plant remains found in ancient Egypt and Mesopotamia indicate that plant cultivation was already widespread by that time. In earlier ruins of pre-Incan indian villages in Peru, archeologists have uncovered Lima beans that have seeds nearly 100 times larger than those of wild Limas in the area. This suggests that the Incans obtained their beans from still earlier plant breeders who left no record.

The second technique was breeding. The farmers selected two plants and then crossed them to produce offspring having the desired traits of both parents. The process was hit or miss, however, since early plant breeders did not understand the genetic transmission of traits and could

5

not predict the likely outcome of a particular cross. Nonetheless, valuable traits did arise that could be selected and maintained in the population.

The physical basis of inheritance—or what actually happens when two plants are crossed—was not understood until the early 1900s. The key was Gregor Mendel's breeding experiments in the 1860s, though the importance of his work was not recognized until after his death. Working with peas in his monastery garden in Austria, Mendel deduced that hereditary information is stored in discrete units that we now call genes. Moreover, he reasoned that each trait, such as color, is controlled by two genes, one from the male parent and one from the female parent.

Soon after, other researchers found that genes are transmitted in blocks of 5,000 or so, rather than independently as Mendel had surmised. What Mendel did not know was that genes do not exist separately in the cell; rather, they are linked together on long chromosomes in the cell nucleus. Thus, while the gene is the unit of heredity, the chromosome is the unit of transmission. Each parent contributes half of the chromosome complement to the offspring; in humans, for instance, each parent contributes 23 chromosomes.

In the early 1900s, biologists learned how chromosomes are assorted during cell division—and how that determines the properties of the offspring. They learned how to locate genes on chromosomes, because chromosomes break and rejoin, or cross over, fairly regularly during cell division, leading to new genetic combinations. They also learned that sometimes chromosomes are present in multiple copies, or in reduced number, and that this particular dosage affects gene expression.

### The First Biological Revolution

The foundation of Mendelian genetics enabled plant breeders to cross plants with new precision, carefully manipulating the plant genome to produce new, improved varieties. These breeding techniques have been used to develop higher-yielding varieties, including plants resistant to pests or disease. These improved varieties have contributed to a dramatic explosion in agricultural output. In the past 50 years in the United States, farm productivity has increased two-and-a-half times, while farm acreage has declined 6 percent. One of the most spectacular successes was the development of hybrid corn in the 1930s, which quickly doubled corn yields.

Breeding advances have also meant more food for the rest of the world. In the 1950s and 1960s, Norman Borlaug at the Center for Maize and Wheat Improvement in Mexico developed semidwarf wheat varieties, and the International Rice Research Institute in the Philippines developed similar improved rice varieties. When introduced in the 1960s in India and later in China, the wheat and rice varieties became the basis of the "Green Revolution," in which crop yields increased an estimated four to seven times. For these reasons, the introduction of applied genetics to agriculture is sometimes called the first biological revolution.

Nonetheless, these productivity gains are not due to genetic advances alone. In the United States, half of this gain is generally attributed to the simultaneous improvements in farm management—in cropping practices, in farm machinery, and especially in the development of new agricultural chemicals such as pesticides, fertilizers, and herbicides. Similarly, the introduction of improved wheat and rice varieties in South Asia was accompanied by a heavy investment in irrigation and agricultural chemicals.

Though agriculture has profited immensely from the improved breeding practices developed from Mendelian genetics, the technology does have its limitations. One problem, as Bogorad described at the convocation, is time. It may take generations and generations to develop a desired strain through selection and breeding. The greatest limitation, however, is simply the available supply of genetic diversity.

Courtesy of U.S. Agency for International Development.

As Darwin discovered more than 100 years ago, new species evolve through natural selection. If part of a breeding population becomes isolated, its gene pool becomes more and more distinct from that of the parent population. Often, biological barriers arise that prevent the two populations from interbreeding. Consequently, within each distinct species, genetic variation among individuals decreases. Because of such natural breeding barriers, the plant breeder in search of useful new variants is confined to members of the same species or closely related species. Compounding the problem, many major crops have been under cultivation for thousands of years, which has led to an increasingly homogeneous gene pool. In some cases, the desired trait is simply not available in the breeding population. "You can breed and breed and breed, and never get the trait you are looking for," Bogorad said.

For this reason, in many essential crops we may be reaching what Lowell N. Lewis, director of the California Agricultural Experiment Station, called a "biological roadblock" in the drive for greater productivity. "Yields have started to level off, and in some cases are declining. For these crops, it is no longer simply a matter of sprinkling on a little more fertilizer."

To Vernon Ruttan, an agricultural economist at the University of Minnesota, the closest analogy to the current situation is the closing of the land frontier in the United States in the 1890s. As land became scarce and expensive, farmers could no longer increase output by simply extending their existing techniques to new land. Instead, increased agricultural output became dependent upon improved varieties and agricultural chemicals, which came into use over the years through the work of plant breeders and agricultural scientists. In essence, Ruttan said, these agricultural chemicals became a substitute for land. Now farmers in the more developed countries are beginning to exhaust the potential of these chemical technologies as well. For instance, the application of nitrogen fertilizer once assured a sizeable boost in yield. Now the gains come harder. Corn is one of many examples. From 1954 to 1960, the use of nitrogen fertilizer increased corn yields by two bushels per acre per year. From 1971 to 1980, fertilizer added only half a bushel.

### Burgeoning Demand for Food

Meanwhile, as the increase in agricultural productivity slows, the demand for food continues to rise. The United States still produces a surplus of grain, but as Orville Bentley, assistant secretary for science and education of the U.S. Department of Agriculture, stated, "for countries that can produce in excess of their needs, there are many more

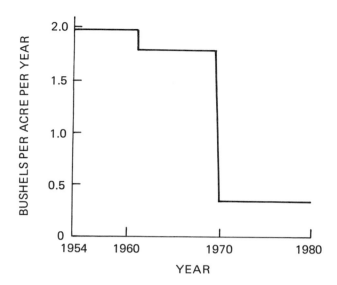

The contribution of nitrogen fertilizer to U.S. corn yields during the periods 1954-1960, 1961-1970, and 1971-1980. From W. B. Sunderquist et al., "A Technology Assessment of Commercial Corn Production in the United States, " Minnesota Agricultural Experiment Station Bulletin 546, 1981.

that are experiencing food shortages." According to W. David Hopper, vice president of the South Asia Region of the World Bank, although there has not been as massive a famine as the one in India in 1943, hunger and malnourishment are still pervasive worldwide. The Food and Agriculture Organization of the United Nations estimates that some 500 million people are severely undernourished.

If world population continues to grow at 1.8 percent annually, food production will have to at least double in the next 40 years to keep pace with demand. Hopper suspects that the demand for food will double in 30 years; as people become more affluent, they will seek a greater and more varied diet. Moreover, said Bentley, "not only will an increasing number of people need to be fed, but that food must be produced from inferior soil under poor climatic and deteriorating biological conditions." Existing biological and chemical technologies may not be adequate for the task.

In the short term, for the next 10 years, Hopper predicted that world food production can keep pace with demand if there is a substantial investment in these biological and chemical technologies. In India and China, for instance, improved varieties of wheat and rice are fairly well distributed. These nations, like other developing nations, now need what Hopper called a steady accumulation of the "betters"—better use of irrigation, better use of pesticides and fertilizer, better agronomic practices. If both improved biological varieties and supporting technologies are provided, these nations should experience a surge in productivity. But within 20 to 30 years, Hopper warned, the developing nations will also begin to exhaust the potential of these technologies. Unless new biological materials—new varieties—are introduced, there will not be enough food for the world's population.

---

## THE THIRD WORLD

Faced with growing populations, many Third World countries urgently need to increase their agricultural output. Yet, introducing new agricultural practices is not a simple task, according to W. David Hopper, vice president of the South Asia Region of the World Bank. Moreover, increased yields will not come simply from introducing improved varieties or cropping practices. There must also be social and economic incentives for the farmer to adopt these new agricultural technologies. In short, the new agricultural technologies must be profitable, even for farmers who practice collective agriculture. There must also be an organizational structure in the country that will support the adoption of new practices. For example, there must be a source of fertilizers, pesticides, and farm equipment, and irrigation must be available. The farmer must also have a market where he can sell his product. And there must be a transportation system linking all of these.

In the 1950s, international development agencies tended to neglect one or more of these components, Hopper said. Yet all three—the agricultural technologies, economic incentives, and infrastructure—coalesced in the 1960s in India and China, Hopper said. Norman Borlaug's improved wheat varieties were introduced, as were the advanced rice varieties of the International Rice Research Institute. These were accompanied by a major investment by the development agencies that allowed the expansion of irrigation systems and the widespread use of agricultural chemicals. The result was the "Green Revolution."

The rest of the world has not been so fortunate, Hopper said. Many African nations, for example, are still "desperately short" of techniques for working with their soils. Genetic engineering can play a major role in developing new varieties suited for these conditions and could direct the future course of agriculture in developing nations. Yet attention must also be given to the supporting technologies that will make these new varieties more productive than the traditional techniques.

Courtesy of U.S. Agency for International Development.

Working with existing gene pools, plant breeders must somehow develop varieties that are higher yielding, more nutritious, adapted to harsh environments, less costly to farm, and perhaps resistant to pests and disease. That is where molecular biology and genetic engineering hold great promise.

## Molecular Genetics

Genetic engineering enables molecular biologists to reshuffle genes in combinations not possible in nature, opening up a vast new source of genetic diversity for crop improvement. "One of the most remarkable achievements of genetic engineering and molecular biology is that we now operationally have a kind of world gene pool," Bogorad explained. "Darwin aside, speciation aside, we can now envision moving any gene, in principle at least, out of any organism and into any organism."

In some cases, gene transfer will entail combining the genes of two plants, as do today's plant breeders—but without the limitations of working with the whole plant. Although Mendelian genetics eliminated much of the guesswork in classical breeding, there is still an element of trial and error: when two entire genomes are combined in a sexual cross, the breeder cannot be certain of the outcome. He may be breeding for one trait, controlled by one gene, but the hundreds of thousands of other genes in each plant complicate the task. By contrast, the molecular genetic engineer can pluck that single gene from the donor plant and insert it into the recipient, leaving the extraneous genes behind. That specificity also brings a saving in time. Through gene transfer, an improved variety can be created in a single experiment, in one generation. Yet, using conventional techniques, it takes repeated back crosses to eliminate the unnecessary genes and thus many generations and several years to create an improved variety.

Moreover, the genetic engineer in search of a gene for pest resistance, heat tolerance, or another trait is no longer constrained by the natural breeding barriers—he can select from any species. Eventually, the genetic engineer may also select from outside the plant kingdom, borrowing genes from animals or bacteria. A recent experiment demonstrated that such transfers are indeed possible: a gene for antibiotic resistance was transferred from a bacterium into a petunia plant, where it conferred resistance on the plant. Eventually, it may be possible to transfer the genes for nitrogen fixation from bacteria to plants, thereby reducing the need for fertilizer.

## Uncertain Impact

Molecular genetic engineering is still in its infancy. It is too early to gauge the impact it will have on agriculture and crop improvement. As Ruttan explained, "It took 30 years to make the transition from getting most of our productivity growth by bringing new land into production to beginning to get it from the old biological technology—the first biological revolution. The question of whether and when the second biological or biotechnological revolution will reverse the current productivity decline is still unanswered."

Some of the simpler new techniques, based on the ability to regenerate plants from cells in culture, are already offering a shortcut in selection and breeding for some plants. These techniques are generally known as somatic cell genetics, as they involve the manipulation of cells, as opposed to genes or whole plants, for crop improvement.

Gene-transfer techniques are far less accessible than somatic cell genetics. Their successful application will depend upon breakthroughs in the understanding of gene expression and regulation, as well as increased knowledge of plant physiology, biochemistry, and development. It is also too early to judge how plants will respond to such manipulation. For those reasons, Ruttan and others have predicted that, even with the much-needed increase in research, the impact of biotechnology will be small until the late 1990s.

Though their specific applications cannot be predicted, these new genetic engineering techniques seem likely to become powerful adjuncts to conventional breeding practices. Ultimately, their success will depend on how well they can be integrated with conventional technologies. Molecular biologists will need to work closely with plant breeders to identify promising projects for genetic engineering. When a new variety is developed in the laboratory, it will face the same scrutiny as does any new variety; it will need to undergo lengthy evaluation in the field. It must perform, offering an advantage in quality, yield, time, or cost, if the farmer is to adopt it. For many major crops, sophisticated and effective breeding strategies already exist; it is unlikely that these new gene-transfer techniques will supplant them. Instead, they may offer the greatest advantage in engineering of crops that are difficult to manipulate by conventional techniques.

The new genetic technologies will undoubtedly aid agriculture in ways that cannot be anticipated now. Cell culture techniques, for instance, are providing a valuable supply of genetic diversity that was totally unexpected when work began a few years ago.

The greatest impact of these new technologies, however, may be in

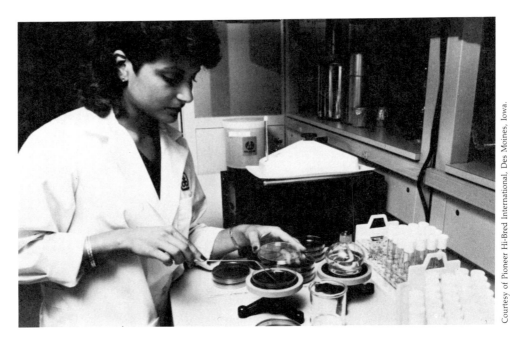

elucidating the basic biology of plants. Though this work is just beginning, gene-transfer techniques are already proving an invaluable tool for exploring the structure, function, and control of genes. This new knowledge can then be used to devise more direct, and thus quicker, breeding strategies—either at the whole plant, cellular, or molecular level.

The vast potential of genetic engineering does not diminish the need for other advanced research. As Lowell N. Lewis pointed out, to increase agricultural output in developing nations will require more research on the biology and ecology of tropical food plants, as well as the pests and diseases that plague them. It will entail bringing underutilized plants into production and a continued search for new, valuable germ plasm.

# Gene Transfer

## The Background

Most of the current excitement in agricultural research focuses on gene-splicing or recombinant DNA technology. Lawrence Bogorad described the development of this technology, its promise, and limitations.

This technique had its impetus in research in the 1940s and 1950s on the molecular structure and function of genes. In the 1940s, Oswald Avery, Colin MacLeod, and Maclyn McCarty presented evidence that genes were made of deoxyribonucleic acid, or DNA. DNA is a molecule consisting of sugar, phosphate, and four bases: adenine, guanine, thymine, and cytosine (A, G, T, and C). At that time, no one could fathom how such a simple molecule could contain and transmit hereditary information.

Watson and Crick provided the answer in 1953. They described DNA as a two-stranded molecule, coiled in the now-famous double helix. The backbone of the molecule is a string of sugar and phosphate. A nucleotide base—either an A, G, T, or C—sticks out from each of the sugars. The two strands are held together by weak bonds between these bases; A binds with T, and G binds with C. Thus, each strand is complementary to the other.

The elucidation of that structure revealed how DNA passes on instructions from one generation to the next. Prior to cell division, the two strands unwind. Each strand then serves as a template for the faithful replication of another DNA molecule, which is then passed on to progeny.

DNA contains hereditary information—but the major question is how is that information processed and turned into a trait? The answer lies in the "central dogma" of molecular biology—that is, information con-

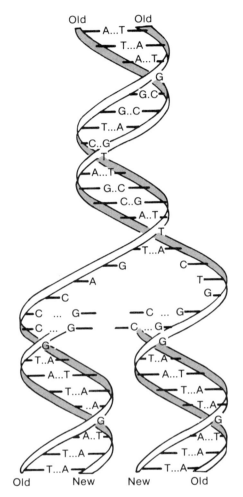

Replication of DNA. The double-stranded DNA helix unwinds and each strand serves as a template for the building of a complementary strand. The resulting daughter DNA molecules are exact copies of the parent, with each double helix having one of the parent strands. From U.S. Congress, Office of Technology Assessment, *Impacts of Applied Genetics: Micro-organisms, Plants, and Animals*, U.S. Government Printing Office, Washington, D.C., 1981.

tained in DNA is copied into molecules known as ribonucleic acid, or RNA, and RNA specifies the synthesis of all proteins. Thus, DNA carries the instructions for proteins—both for the structural proteins, such as those in the framework of membranes, and for the enzymes, which catalyze all the metabolic reactions of an organism.

Proteins are chainlike molecules composed of a sequence of amino acids. That is where the genetic code comes in: a series of three nucleotide bases in DNA codes for each amino acid in a protein. The sequence TAC, for instance, codes for the amino acid methionine; the sequence TAT codes for the amino acid isoleucine. A single change in a nucleotide means that one amino acid in the protein is replaced by another. "That's a mutation in the gene and an alteration in the gene product," Bogorad explained. "It came as a surprise 20 years ago when it was discovered that a single amino acid change in a protein could greatly affect the way it worked."

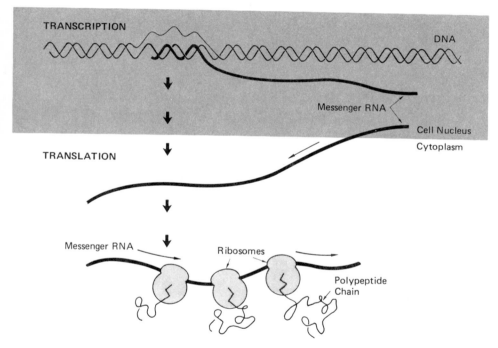

The "central dogma." DNA carries the instructions for the synthesis of proteins. A series of three nucleotide bases in the DNA molecule code for a specific amino acid, the building blocks of protein. Each gene, a relatively short segment of a long DNA molecule, codes for a single protein. The genetic information is expressed in a two-step process, described by the "central dogma" of molecular genetics: DNA is transcribed to RNA, then RNA is translated to protein. During transcription, a strand of DNA serves as a template for the formation of a complementary strand of messenger RNA. Next, the messenger RNA moves from the cell nucleus to the cytoplasm. There ribosomes attach to the messenger RNA and direct protein synthesis by reading the genetic code and building a chain of amino acids.

The conversion of a gene to RNA and then to a protein product is called *expression*. If a gene is present and its protein product appears, the gene is said to be "on." If the gene is present but no product appears, it is said to be "off." Gene expression is a two-step process: first, the DNA is *transcribed* to RNA; then the RNA is *translated* to protein.

Transcription is similar to DNA replication. The DNA molecule unwinds, but in this case the strands serve as a template for the formation of an RNA molecule. RNA contains three of the same nucleotide bases as does DNA—A, G, and C. But in place of thymine, RNA has uracil, U. Thus, during transcription, G binds with C, and now A binds with U. Translation occurs when the RNA molecule, known as messenger RNA, leaves the nucleus and travels to the ribosomes, the site of protein synthesis in the cytoplasm. Here the RNA specifies the sequence of amino acids in a protein according to the triplet codes mentioned earlier.

**The Technique**

In theory, gene-splicing is relatively straightforward. In practice, it is far from routine. Bogorad outlined the basic procedure. The first step is to locate the desired gene among the 5 million or so in the cell nucleus. Each gene has three regions, all essential for successful functioning. The beginning is the promoter region, the series of nucleotides that is recognized by the enzyme that triggers the transcription of DNA to RNA. The middle sequence of nucleotides contains the code, the instructions for producing a specific protein. The end series of nucleotides, or terminator, is a signal to stop the transcription process.

Next, the gene must be isolated from the others on the chromosome. For this task, the genetic engineer uses a restriction enzyme. These enzymes recognize specific nucleotide sequences and cut the DNA at precisely those points. It is these enzymes that allow researchers to snip a gene out of the DNA sequence from one organism and splice it into the DNA of another. In fact, the advent of recombinant DNA technology can be traced to the discovery of these restriction enzymes in the early 1970s. Now, a decade later, biological suppliers offer hundreds of restriction enzymes for sale, each one recognizing a different sequence of nucleotides.

Once the gene is isolated, it must be cloned, or duplicated, and inserted into the host cell. Both steps are accomplished by inserting the gene into a plasmid. A plasmid is a tiny, circular piece of bacterial DNA. Plasmids reside as separate units of DNA inside the cytoplasm of a bacterial cell. With the same restriction enzyme that was used to excise the gene from the donor cell, the genetic engineer cuts open the plasmid.

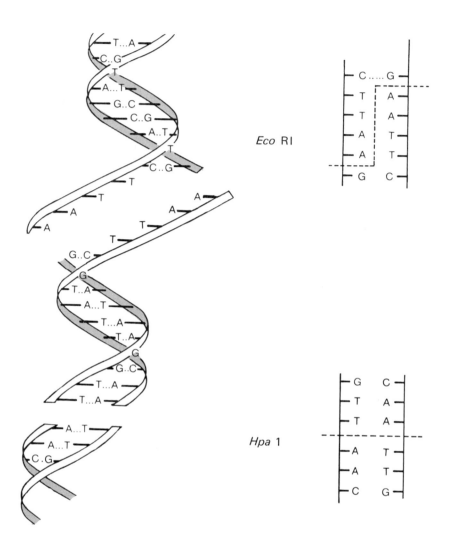

Restriction enzymes. Restriction endonucleoses are enzymes that cleave DNA at specific sites. The illustration shows examples of cleavages made by two such enzymes, *Eco* RI and *Hpa* 1. *Eco* RI recognizes the DNA sequence $\frac{\text{CTTAAG}}{\text{GAATTC}}$ and cleaves each strand between the G and A yielding single strand ends. Such "sticky ends" can readily join on to other DNA fragments created by the same enzyme. *Hpa* 1 does not leave "sticky ends"; it recognizes the sequence $\frac{\text{CAATTG}}{\text{GTTAAC}}$ and cleaves each strand between the A and T. Restriction enzymes can be used to isolate single genes.

This leaves the plasmid with two "sticky" ends, which will now accept the foreign gene. After the foreign gene is inserted, another enzyme, called a ligase, is used to sew the plasmid together. Plasmids are ideal vectors for carrying the new gene into a host cell, because, in nature, plasmids are routinely passed from one bacterium to another, where they are readily accepted. When the recombinant molecule—part bacterial plasmid, part plant gene—is taken up by the bacterial cell, that cell is said to be *transformed*. As the plasmid replicates inside the host cell, it copies the foreign gene along with its standard gene allotment.

The goal is not just to clone the gene, but to have that gene expressed—to have the DNA transcribed to RNA, and the RNA translated into the desired protein—in the host cell. Most work to date has involved the insertion of a gene from a higher organism—usually an animal—into a bacterial host, where the animal gene produces proteins such as insulin, interferon, and human growth hormone. The genes of higher organisms (eukaryotes) have different control signals—signals that turn genes on and off—than do the genes of primitive organisms (prokaryotes) like bacteria, which lack a nucleus. These signals must be read correctly for the gene to be expressed. Gene expression has been achieved by removing the specific control signals from genes of higher organisms, in essence tricking the bacterium into accepting the foreign gene as a bacterial gene.

Up to this point, the techniques for plant genetic engineering are similar to those used to design bacteria to produce insulin or other pharmaceuticals. In short, a gene is isolated, spliced into a vector, inserted into a host cell, and expressed. Pharmaceutical applications depend upon a method of culturing large batches of these recombinant bacteria. By contrast, plant genetic engineering depends upon a means of regenerating a whole plant from the cells in culture.

There are three tissue-culture techniques for regenerating plants from culture (*see* Cell Culture, p. 34). For gene-transfer experiments, the preferred route is the culture of protoplasts—single cells from which the cellulose wall has been removed. That is because it is easier to insert genes into protoplasts than into cells containing the tough outer wall, which animal cells lack.

Thus, before a foreign gene is introduced into a plant cell, that cell is treated with enzymes that dissolve the outer wall. The protoplast, containing its new gene, is then placed in a broth of plant hormones and nutrients that induce it to regenerate. It first re-forms a cell wall. By changing the nutrient mixture, the cells can be induced to multiply and form embryolike structures. Known as somatic embryos, these give rise to tiny plants, which then can be transferred to the soil.

## Current Constraints

Though major strides have been made in the past few years, only the barest beginnings have been made in the transfer of genes among higher plants. As Bogorad explained, the major limitation is the lack of knowledge about basic plant biology necessary to exploit this new technology.

Each step of the process presents its own difficulties. For instance, just finding the desired gene is a monumental task—"one of the most difficult and challenging operations in molecular biology," Bogorad said. The plant genome is large and exceedingly complex. Some genes are located on chromosomes in the cell nucleus. Others are contained in two organelles—the chloroplasts and mitrochondria. Similarly, there are difficulties in identifying all the important parts of the gene, including any DNA sequences necessary to regulate expression of the gene; in developing an appropriate vector to carry the foreign gene into the plant cell; and, finally, in regenerating plants from the transformed cells in culture.

### Vectors

One of the major challenges is the development of vectors to ferry foreign DNA into the plant genome. Only a few bacterial plasmids will work in plants. One of these is the Ti plasmid from the soil-borne bacterium *Agrobacterium tumefaciens*. It is the most promising vector to date for plant genetic engineering. *Agrobacterium* causes crown gall disease: it infects the plant stem tissues, inducing tumors. The disease-causing agent is the bacterium's Ti (for tumor-inducing) plasmid. This plasmid does its damage by inserting itself into the plant cell's genome, where it is replicated and expressed along with the plant's DNA. The expression of the bacterial Ti plasmid genes causes the abnormal cell growth characteristic of crown gall disease. Actually, only a small piece of the Ti plasmid is inserted into the plant genome—this piece is called T DNA (for transferred DNA).

The Ti plasmid is a natural vector that routinely inserts new DNA into plant cells. Moreover, it comes equipped with a trait molecular biologists were seeking: its genes can be expressed in the environment of the plant genome; the regulatory signals of the bacterial genes can be read by the plant cell. Several scientists reasoned that the Ti plasmid could be tricked into carrying additional genes into the plant genome as well.

For the past several years, there has been an intensive research effort to develop the Ti plasmid as a genetic engineering vector. Much of it

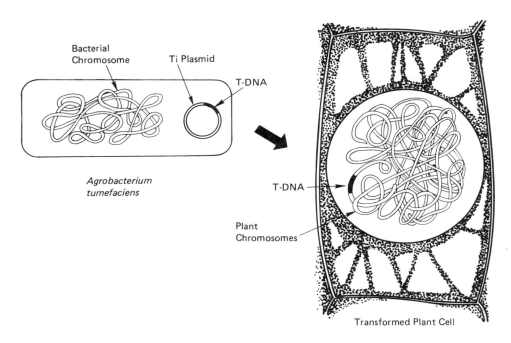

The Ti plasmid vector. In plant genetic engineering, the Ti plasmid can be used to carry foreign genes into plant cells. The Ti plasmid is the disease-causing agent of the soil-borne bacteria *Agrobacterium tumefaciens*. When the bacteria infect a plant, a part of the Ti plasmid called the T DNA is transferred to a plant chromosome. When the T DNA is expressed as part of that chromosome, it causes the plant cell to divide and grow abnormally. Researchers have recently developed procedures for removing the tumor-causing genes from the T DNA and replacing them with desirable genes. The Ti plasmid containing the altered T DNA region can then be used to insert the desired genes into plant chromosomes.

has been performed by two research groups: one led by Mary Dell Chilton at Washington University and the other a European group led by Jeff Schell of the Max Planck Plant Breeding Institute in Cologne, Germany, and Mark Van Montagu of State University in Ghent, Belgium. They have addressed a number of questions, such as how to insert new genes into the T DNA region without disrupting the sequences that control its insertion into the plant genome and how to remove the disease-causing part of the plasmid so that the vector could be used in practical as well as experimental gene transfer.

Their efforts have paid off. In January 1983 the European researchers and another group at Monsanto Co. announced that they had used a Ti plasmid to carry a functioning bacterial gene into a plant cell. This was the first demonstration that a foreign gene could be inserted into a plant cell and be expressed. The Monsanto group included Robert Horsch, Stephen G. Rogers, and Robert T. Fraley. Both teams, working independently, inserted a bacterial gene for antibiotic resistance into the T DNA portion of a Ti plasmid. The Ti plasmid was then used to transform petunia cells in culture. The foreign gene was expressed: the cells in culture were resistant to the antibiotic. When a plant was regenerated from these cells, it retained the antibiotic resistance.

Commenting on the widely heralded gene transfer, Robert M. Goodman of Calgene, Inc., interjected a note of caution. "Notwithstanding the excitement generated by the recent demonstration that the expected is possible, we are only at the beginning of a long period of research and development" on vectors. At this stage, biologists do not understand how the Ti plasmid works—specifically, they do not understand the signals that control the insertion and expression of T DNA.

Even without that knowledge, the Ti plasmid can be effectively used as a vector, as shown by the recent petunia experiment. But Goodman cautioned against becoming too intent on applications, on getting "too carried away with the short-term excitement that we overlook the need to invest in work that leads to an understanding of the underlying principles . . . that will allow the design of sophisticated genetic engineering vectors." For example, he said, "we must understand how the T DNA inserts if we ever hope to control the location and perhaps the multiplicity of the insertion."

According to Goodman and others, additional work is necessary to improve the efficiency of Ti plasmid as a vector. In most work to date, only a small percentage of plant cells inoculated with *Agrobacterium* carrying the Ti plasmid are transformed. If the Ti plasmid is to be used in a practical gene-transfer system, then much higher rates of transformation must be achieved. Such research has already begun.

Other vectors will also be necessary. A major limitation of the Ti plasmid is that it works only in those plants that the *Agrobacterium* normally infects. It does not infect plants in the grass family, which include the important cereal crops like corn, rice, and wheat. Consequently, there is now no vector available for their genetic engineering. It may be possible to modify the Ti plasmid so that it can infect grasses, yet other vectors having different host ranges should be explored.

Another limitation is that the Ti plasmid can be used only to ferry genes into the nuclear DNA. Several agronomically useful traits, however, are controlled by genes located outside the nucleus in either the chloroplasts or mitochondria. For example, male sterility in several cases is affected by genes in the mitochondria. Male sterility is a desired trait in plant breeding because it allows the inexpensive production of hybrid seed. Yet at present, no vector is available to manipulate genes in the organelles.

Plant viruses are also being studied as possible vectors. Viruses are tiny bundles of either DNA or RNA encased in a protein coat. Viruses are of interest because they somehow command the plant cell's machinery to replicate the virus and express the viral genes. As with a plasmid, the idea is to insert foreign DNA into the virus and use it to transform a plant cell. Again, the virus would have to be "disarmed" so that it would not cause disease before it could be used as a practical vector. One advantage viral vectors offer is that almost all plants are susceptible to one virus or another. By contrast, the host range of the Ti plasmid is quite limited.

Most work has been performed on the cauliflower mosaic virus, a small, double-stranded DNA virus. Researchers have inserted short pieces of foreign DNA into the virus. This virus has then been used to transform plant cells. In these experiments, the foreign DNA has been replicated inside the plant cell as part of the viral genome.

But much work needs to be done to develop the cauliflower mosaic virus as a vector. Its major drawback is the size of the foreign DNA it can accept. So far, only pieces of DNA smaller than a gene have been stably inserted into the virus genome. It seems that the protein coat on the virus could be the limiting factor. But there are other plant viruses. RNA viruses, for example, can function without their protein coats and thus conceivably could accept larger pieces containing foreign genes.

Viroids are another possibility, though their use is even more speculative, both because of their size and because so little is known about them. Viroids consist of RNA without the protein coat. They are the smallest pathogenic agents known, smaller than the average gene. The potato spindle viroid, for example, contains only 359 nucleotide base pairs, as opposed to about 8,000 in the cauliflower mosaic virus.

## Gene Expression

Until recently, one of the key uncertainties of genetic engineering was whether a foreign gene would be correctly expressed in a higher organism. The recent successful transfer and expression of a foreign gene into a plant cell answered that question. But, as Bogorad pointed out, achieving expression is only the first step. The next step is to control that expression so that the gene is switched on in the right place at the right time. That selective expression is what occurs in nature: although all cells contain the genes for photosynthesis, for instance, they are turned on only in the leaf, not the root. Molecular biologists have yet to master those controls. Until they do, a transferred gene could simply be on all the time in all the cells.

In numerous laboratories, biologists are searching for the control elements that regulate gene expression. If these controls can be identified and transferred to a plant along with the desired gene, they will permit gene expression to be "targeted" to specific organs and developmental stages. The regulation of gene expression is also of enormous theoretical interest, for the switching on and off of genes determines how a single cell differentiates into a plant.

Another question is whether the introduction of a foreign gene will affect the expression of the other genes. The introduction of new genes through conventional plant breeding can have deleterious effects, suggesting that gene interaction is quite complex. For instance, in 1964 a strain of high-lysine corn was identified. Lysine is an essential amino acid in the diet of nonruminant animals. Though the strain has improved nutritional value for swine and poultry, it is not grown commercially because the yield is reduced 10 percent over other strains. Another drawback is that the kernels of the high-lysine strain do not have good storage quality. It is too soon to say whether molecular genetic engineering will involve similar trade-offs.

## Single and Multigene Traits

Based on the recent advances in identifying and isolating genes as well as advances with vectors, many molecular biologists are confident that they will be able to engineer traits controlled by a single gene or a small cluster of genes. Yet many commercially valuable traits, such as yield and stress resistance of various kinds, are controlled by numerous genes somehow acting in concert. These are called multigene traits. Just finding the genes will be difficult, as evidence suggests that they are scattered throughout the chromosome. Determining how the expression of these genes is regulated, and how their gene products interact, is an even more formidable task. Consequently, it is not clear that techniques can be developed to engineer multigene traits.

## Plant Regeneration

Successful gene transfer ultimately depends on the ability to regenerate plants from cells in culture. Yet protoplast culture is far from a proven technology. Although it works well in some species, such as carrots, tomatoes, tobacco, and petunia, some of the major crop species are notoriously difficult to regenerate from protoplasts. Potatoes and alfalfa have been successfully regenerated from protoplasts, but the technique does not work reliably for corn, wheat, or soybeans. Compounding the difficulty, no one knows exactly why.

The secret lies in the signals that turn genes on and off during development. Cell culture is an attempt to mimic that process in the laboratory. As the ability of plants to regenerate reveals, cells are totipotent—that is, each cell, such as a leaf cell, contains the instructions for the whole plant. In the differentiated state—in the leaf, for example—most of those genes are shut off. The trick is to induce the cell in culture to regress to an undifferentiated state in which the genes can be switched on and off again in proper sequence.

The development of culture methods has been handicapped by the lack of knowledge about the regulation of gene expression during development. Research to understand the genetic mechanisms involved in regeneration is proceeding in tandem with efforts to develop practical culture techniques. Although an increasing number of plants are yielding to protoplast and the other culture techniques, these advances have stemmed as much from guesswork as from science. Whether or not a plant will respond in culture is influenced by several factors, including the composition of the nutrient broth, the specific genotype of the donor plant, and the site from which the explant is taken. In working with unresponsive species, biologists are often confined to juggling these factors—perhaps screening hundreds of genotypes in search of one that will work. To a lesser degree, similar uncertainties surround the other two in vitro regeneration techniques: callus and suspension culture (see Cell Culture, p. 34).

# A Tool for
# Fundamental Plant Science

Molecular biology and genetic engineering are already having a major impact on fundamental plant science. Gene cloning, gene transfer, and other new techniques are proving valuable research tools for probing gene structure, function, and plant development. This knowledge, in turn, can be used in designing increasingly sophisticated methods to engineer improved crops.

Because of these recent advances, molecular biologists now know as much about the structure, organization, and expression of the plant genome as they do about the animal genome, according to Robert Goldberg of the University of California at Los Angeles. Nonetheless, in both areas, there is still much to learn. Far more is known about bacterial genes than the genes of either plants or animals. Bacteria are less complex than higher organisms and comparatively easier to study. A single-celled bacterium contains some 5,000 genes, whereas the average plant genome is roughly 1,000 times larger, about the same size as the human genome. (The actual size of plant genomes vary greatly in size from one taxonomic group to the next.) To date, only about a dozen plant genes have been fully sequenced.

As Goldberg described, molecular biologists are learning that the organization of the plant genome is also exceedingly complex. Certain sequences of the DNA are repeated hundreds to thousands of times. No one knows the function of these reiterated sequences.

Interspersed among the coding regions, plant genes also contain sequences of DNA that do not code for protein at all. These noncoding sequences are called introns to distinguish them from the coding regions, called exons. "These introns seem to have 'popped' into the gene during evolutionary time, splitting the regions that code for proteins," Goldberg explained. Introns interrupt the genetic code—the instructions for pro-

27

## SIZE OF PLANT GENOMES

| | $5 \times 10^2$ | $5 \times 10^3$ | $5 \times 10^4$ | $5 \times 10^5$ | $5 \times 10^6$ | $5 \times 10^7$ kb[a] |
|---|---|---|---|---|---|---|
| Flowering Plants | | | | ▨▨ | ▨▨▨ | ▨▨ |
| Birds | | | | ▨ | | |
| Mammals | | | | | ▨ | |
| Reptiles | | | | | ▨ | |
| Amphibians | | | | | ▨▨▨ | ▨ |
| Bony Fish | | | | ▨ | ▨ | |
| Cartilaginous Fish | | | | | ▨ | |
| Echinoderms | | | | ▨ | ▨ | |
| Crustaceans | | | | | ▨ | |
| Insects | | | | ▨▨ | ▨ | |
| Mollusks | | | | ▨ | ▨ | |
| Worms | | | ▨ | | | |
| Molds | | | ▨ | | | |
| Algae | | | ▨ | | | |
| Fungi | | ▨ | | | | |
| Gram-positive bacteria | | ▨▨ | | | | |
| Gram-negative bacteria | | ▨ | | | | |
| Mycoplasma | ▨ | | | | | |

[a]kb = 1,000 base pairs.

|  | $10^3$ | $10^4$ | $10^5$ | $10^6$ | $10^7$ | $10^8$ kb |

A comparison of the size of genomes. Plant genomes are 100 to 10,000 times larger than bacterial genomes. Courtesy of Robert Goldberg, Department of Biology, University of California at Los Angeles.

teins—and must somehow be removed before the gene can be expressed. Molecular biologists have learned that genes possess their own mechanism for excising the introns; it is remarkably similar to methods developed by genetic engineers. The DNA is transcribed to RNA, then in a series of enzymatic cut-and-paste reactions, the introns are excised from the RNA. This processed strand of messenger RNA, minus the introns, can then be translated to protein. Because no one has yet demonstrated a function for introns, they have been jokingly referred to as junk DNA. If, however, they prove to have a role in gene expression, introns will present additional complexity and perhaps opportunities in genetic engineering.

Molecular biologists can determine how many genes are actually ex-

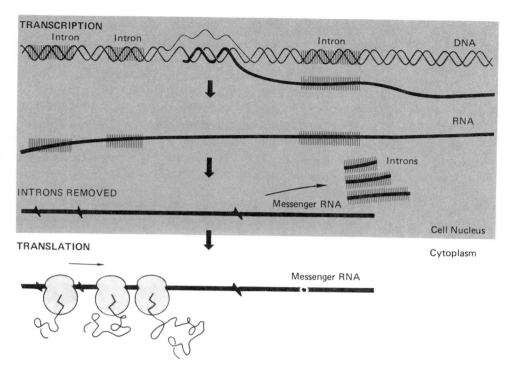

Removal of introns. In plants and animals the DNA sequence of a gene that codes for a protein may be interrupted by noncoding stretches called introns. These introns must be removed before the gene can be expressed. The cells have a natural mechanism for excising introns. First, the DNA containing the introns is transcribed to RNA. Then enzymes within the cell nucleus cut the introns from the RNA and splice the coding sequences back together. The resulting messenger RNA then migrates to the cytoplasm to be translated into protein.

pressed—actively making protein—at a given time in the life cycle of a plant. This is easily done because RNA can be used to identify the DNA from which it was transcribed. In tobacco, Goldberg reported, about 100,000 genes are active at a specific time in the life cycle of the plant. In other words, only about 5 percent of the DNA contained in the nucleus is used at one time to produce proteins. The function of the other 95 percent is unknown, though regulatory sequences are known to account for some of this DNA.

The next question, Goldberg said, is the extent to which these genes are regulated in the plant: "Do all these organ systems have 100,000 genes that are on and active, or is there a differential gene expression that could contribute to the differentiated state of the particular cell?" In the tobacco plant, approximately 25,000 genes are on in each organ

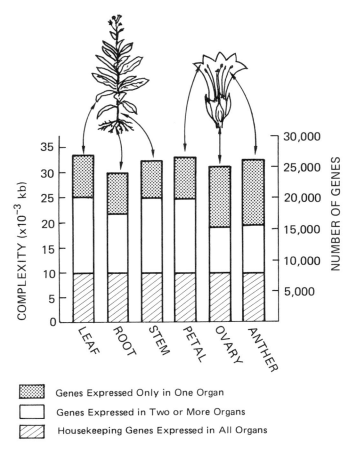

Gene expression is under complex control in higher plants. Some genes are only expressed in specific parts of the plant, while other genes are expressed in all cells of the plant. Courtesy of Robert Goldberg, Department of Biology, University of California at Los Angeles.

system, e.g., in the leaves, stems, or flowers. The most important finding, according to Goldberg, is that each organ system has a unique set of genes—genes that are expressed only in that organ. For example, the petals and the leaves each contain approximately 7,000 specific genes. The ovaries and the anther contain approximately 10,000 specific genes. In addition, all the organs share some common genes, which Goldberg calls "housekeeping genes," that code for proteins necessary for all cells in the plant. In short, he said, plant genes are highly regulated.

The regulatory circuitry is proving to be even more intricate than anticipated. Not only can genes be either active or silent, but there now seems to be an intermediate state. Some genes are switched on—they are transcribed into messenger RNA—but for some reason, the messenger RNA is never transported to the cytoplasm to the site of protein synthesis. Since no proteins are made, the genes are effectively off. Thus, in addition to the standard "on-off" switches, molecular biologists are searching for the regulatory mechanism that determines whether an active gene will actually be expressed.

The central task in both molecular and developmental biology is to understand the regulatory controls responsible for differential gene expression during development—to find the DNA sequences that instruct a gene to be on in a leaf but off in a root. Only when these DNA sequences have been identified can molecular biologists use gene-splicing techniques with predictability to engineer new crop plants.

## Seed Protein Genes

Elucidating the mechanism of these regulatory controls is not easy. Goldberg and his colleagues at UCLA are tackling the problem by examining the genes that code for storage seed proteins in the soybean plant. These proteins, contained in most plants, provide a source of nutrition for the plant when the seed germinates and begins growth. They are also an important food source for livestock and humans.

Goldberg is studying the genes for these seed storage proteins for two reasons. First, he said, the genes are active at only a specific time in the plant's life cycle—during embryogenesis. That makes them a useful system for examining the regulation of gene expression. Second, such research could have agricultural significance. Although seed storage proteins in soybeans and other crops are important food sources, they are deficient in some amino acids essential to human nutrition. If that deficiency could be corrected, the nutritional value of these crops would be significantly enhanced. It might be possible to correct that deficiency through genetic engineering.

There are two possible approaches. First, some of the storage proteins in a soybean contain more of the essential amino acids than do the other storage proteins. One approach would be to engineer the plant to produce more of the protein that contains an abundance of essential amino acids. That involves changing gene regulation so that the expression of one gene is favored over the expression of another. The other approach would be to directly engineer the seed storage protein gene—inserting more codes for essential amino acids. Ultimately, molecular biologists may attempt to construct a seed protein gene *de novo*. For any of these approaches, the molecular biologist must know exactly when during development the different storage protein genes are turned on, and where the switches are.

Using molecular techniques, the UCLA researchers have been able to pinpoint exactly when during embryogenesis the seed protein genes are active. They have also confirmed, as expected, that the genes are not active in the adult plant. They have isolated a storage protein gene and are now studying its structure. They have found that it contains three introns, or noncoding sequences, and they have located the regulatory sequences responsible for excising those introns in the natural cut-and-paste reaction. They have also identified the gene's promoters and terminators—the sequences responsible for starting and ending transcription. Yet they have not been able to identify the regulatory sequences responsible for turning the gene on and off during embryogenesis.

# Somatic Cell Genetics

While the practical applications of gene-splicing are yet to be realized, cell culture techniques are already proving a valuable tool in crop improvement. For thousands of years, breeding has been based on genetic diversity and selection of desirable traits. The ability to regenerate plants from cells in culture has given rise to new techniques—generally referred to as somatic cell genetics—that both increase the supply of genetic diversity and make possible more efficient selection.

## Selection

Millions of cells, each a potential plant, are typically grown in a single flask in cell-suspension culture (*see* Cell Culture, p. 34). This offers a tremendous potential to use biochemical agents to identify and select useful variants, saving both the space and the time of screening whole plants in the field. As Stephen P. Baenziger, a plant breeder with the Agricultural Research Service, explained, "When I plant wheat, I plant 1,800,000 plants per acre. When my colleagues do a biochemical selection, they plate between 3 and 5 million cells in a petri dish. That single petri dish with an area of about 6 square inches is the equivalent to one-and-a-half to two-and-a-half acres of wheat plants. If I were a corn breeder, that would be the equivalent of 120 to 200 acres."

In conventional selection, a breeder applies a herbicide, fungal pathogen, or other selective agent directly to plants in the field. In cellular-level selection, the breeder simply douses the cells in culture with the herbicide or other selective agent, screening millions of cells at one time. The resistant cells are those that live. They would then be regenerated to see if the trait is still expressed in the whole plant. If the regenerated

33

plant retains resistance, then its progeny must be evaluated to see if the trait is stably inherited.

Not all traits can be selected as easily as herbicide resistance, where the trait itself is the selective agent. For other characteristics, such as height, there is no direct biochemical assay at the cellular level. Researchers are looking for other biochemical markers that will enable them to select such traits in culture.

---

**CELL CULTURE**

There are three methods for regenerating plants from cells in culture: callus, cell-suspension, and protoplast culture. The most reliable of these is callus culture. Through experiments with agricultural species began just a few years ago, many can now be routinely regenerated from callus culture. In this approach, a tiny piece of tissue is snipped from a seedling shoot or other appropriate plant part and placed in a petri dish containing the plant hormones auxin and cytokinin, along with organic and inorganic nutrients. The cells grow and divide, forming a mound of undifferentiated cells called a callus. When transferred to a regeneration medium, the cells in the callus differentiate into roots and shoots, which then grow into plants.

Since hundreds to thousands of plants can be regenerated from one piece of tissue, callus culture offers a means of cloning far more plants in less time than is possible using conventional vegetative propagation. Indeed, since the 1960s, callus culture has been used in the mass cloning of orchids and other horticultural plants that are difficult or costly to propagate otherwise. It is also a promising technique for propagating trees and other slow-growing species. The drawback is that callus culture is labor-intensive and expensive. For that reason, it is not yet being used commercially for the propagation of any agricultural crops.

Most crop improvement schemes involving genetic engineering hinge on the ability to regenerate plants from single cells, not clumps of tissue. For that, either cell-suspension or protoplast culture is used. In suspension culture, a piece of callus is agitated in a flask containing a liquid medium. The callus breaks apart into single cells or clumps of two or

Corn plants regenerated
from tissue culture.
Courtesy of Calgene, Inc.

more cells. These cells then regenerate either by forming roots and shoots or else by forming somatic embryos, which then differentiate into entire plants.

Achieving regeneration from single cells is far more difficult than starting from a clump of tissue. Many species that readily regenerate from callus lose that ability in suspension culture. In general, the plant family Solanaceae—including petunia, tobacco, and potato—are the most responsive to cell-suspension culture; the cereals and legumes are typically very difficult to regenerate. Recent results have been encouraging: corn can now be regenerated from cell-suspension culture, and the list of species is growing yearly.

Protoplast culture is the regeneration of plants from single cells from which the outer wall has been enzymatically removed. Because protoplasts must be induced to re-form their cell walls and then proceed through callus and somatic embryogenesis, this culture technique is more complicated than the other two. Successes in crop plants are rare—potatoes and alfalfa are notable exceptions—and poorly understood. In some cases, regeneration can be achieved by starting with cells from suspension culture, rather than isolating cells directly from a plant. A concerted research effort is under way, however, because protoplasts are the preferred host cell for gene-transfer experiments.

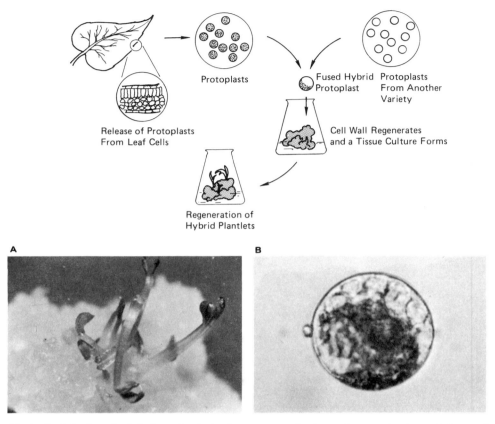

Protoplasts

Release of Protoplasts
From Leaf Cells

Fused Hybrid
Protoplast

Protoplasts
From Another
Variety

Cell Wall Regenerates
and a Tissue Culture Forms

Regeneration of
Hybrid Plantlets

A

B

Protoplast fusion. Isolated protoplasts from some plants such as petunia, potato, and clover can easily be cultured and regenerated into whole plants. This offers the opportunity to fuse protoplasts from different plants to form hybrids that combine their characteristics. Photograph A shows plantlets of a wild species of clover regenerating from a tissue culture derived from an isolated protoplast. Photograph B is a hybrid protoplast created by fusing a protoplast of red clover with one from the wild species. The darker regions of this hybrid protoplast are from red clover. Courtesy of Glenn B. Collins and Jude Grosser, Agronomy Department, University of Kentucky, Lexington.

## Protoplast Fusion

Under appropriate conditions protoplasts from different plants can be induced to fuse in culture, combining their genetic information to create a new hybrid. The recombinant protoplast can then be induced to reform a cell wall, proliferate, form callus, and regenerate. This technique can be used to fuse protoplasts of the same species or of different species. It is the latter possibility that has engendered the most excitement: the creation of entirely new hybrids from species that cannot be crossed sexually. It is also the most speculative. It would combine within one cell wall two complete—and different—sets of developmental instructions, which may be incompatible. Researchers have had some success in fusing protoplasts of different species. Yet to date, only hybrid protoplasts from closely related species have been induced to regenerate from culture. In addition, fusion lacks the precision of gene-splicing, in which a specific gene can be transferred to create a carefully tailored plant. One approach to both of these problems is to delete part of the genome in one of the two protoplasts. For example, cytoplasmic traits— those controlled by the DNA in chloroplasts or mitochondria—could be transferred separately by using a donor protoplast from which the nucleus was removed.

## Somaclonal Variation

Cell culture is also making available a new, unanticipated source of genetic diversity. It was originally assumed that plants regenerated from the same clump of tissue would be identical. Yet many of the plants arising from undifferentiated cells in culture are strikingly different from each other and the parent plant from which the culture was derived.

In some as yet unknown way, the process of culturing cells—of going from a differentiated state to an unorganized state and back to a differentiated state—releases a pool of genetic diversity. William Scowcroft of the Plant Industries Division of CSIRO in Australia likens it to a genetic earthquake moving through the genome, rearranging the genetic information. The exact cause of this somaclonal variation, as it is called, is uncertain, although theories abound. What is clear is that the phenomenon is ubiquitous, occurring in rice, corn, wheat, barley, potato, alfalfa, rape, and other species, and affecting many agronomically useful traits. In several species, for instance, the somaclonal variants include resistance to diseases: sugarcane has developed resistance to eyespot disease, Fiji virus, downey mildew, and smut; potatoes to late and early blight; corn to Southern corn leaf blight; and oil seed rape to vitricular disease.

If it can be harnessed, if useful variants can be selected from culture and used for breeding, this variation could be an unexpected windfall for plant breeders. Scowcroft is optimistic: "I believe that somaclonal variation is accessible—and I hope that with more knowledge it will be manageable. Most certainly, I believe it is applicable for the real world of plant breeding."

Scowcroft and his colleagues are looking for useful variants arising from the culture of wheat. According to Scowcroft, they pay scant at-

Somoclonal variation. Plants regenerated from tissue culture often show dramatic variation in agriculturally important traits. Although the genetic mechanism for this somaclonal variation is not fully understood, it can provide a valuable source of genetic diversity for plant breeding. The photographs show somaclonal variants of wheat. Variation in plant height is compared to parent plants on the extreme right and left. Seed head morphology can differ markedly from the parent type shown in the center of the top photograph. Courtesy of William R. Scowcroft, Division of Plant Industry, CSIRO, Canberra, Australia.

tention to the primary generants—the plants regenerated directly from culture—because much of the variation that occurs in them is unstable. Instead, they look to their progeny to determine if traits are stably transmitted. The stable traits resulting from somaclonal variation could be caused by a variety of mechanisms, including chromosome breakage and reunion, DNA rearrangement, and point mutations—the substitution of a single nucleotide base. At this stage, however, research is just beginning on the causes of somaclonal variation. The amount of variation appears to be affected by several controllable factors, including length of time the cells are in culture, the genotype, the medium, and the culture conditions. An understanding at the molecular level of the factors that control the stability or instability of the plant genome would provide another powerful tool for crop improvement.

Scowcroft and colleagues have tracked useful variants through several generations. Some plants differ in just one trait; others have multiple changes. One of the most unexpected, and welcome, findings is that stable variation occurs in multigene traits, such as height and maturation date, as well as in single-gene traits. In wheat, they have found a variation in height, color, number of side shoots (tillers), and the shape of the awns that surround the grains. Variation also occurs in biochemical characteristics, such as the production of alpha amylase enzyme and in the seed storage proteins. Many of these are potentially useful.

Scowcroft's research group is already attempting to use somaclonal variation in wheat improvement. Specifically, they are screening cells in culture for traits that would be useful in no-till farming, which is becoming increasingly prevalent as a means of conserving soil. Some existing cultivars are poorly suited for no-till farming. Desired new traits include rapid establishment, herbicide tolerance, winter habit, and disease resistance. Somaclonal variation may also provide genotypes suited for tropical environments—genotypes able to tolerate heat or acidic soils containing harmful levels of aluminum and manganese.

# Applying the Tools of Biotechnology to Agricultural Problems

Effort is now under way to apply these cellular and molecular technologies to specific agricultural problems. For instance, researchers are attempting to develop herbicide-resistant plants and plants less susceptible to environmental stresses, such as drought, salty soils, or climatic extremes. They are working on increasing the nutritional quality of feed crops. They are also trying to engineer soil microorganisms that can be used to supply nitrogen to crop plants, or else mitigate or combat soil diseases.

## Herbicide Resistance

Some 420 million pounds of herbicides are used by farmers each year in the United States. Herbicides are used to kill the weeds that compete with crop plants. Unfortunately, they can also kill some crop plants as well. This restricts each herbicide to use on a resistant crop. Nevertheless, crop losses from herbicides can still occur. A case in point is the herbicide atrazine, which is commonly used in the culture of corn. In Illinois, for instance, 90 percent of corn acreage is treated with atrazine each year. Corn can tolerate atrazine because the plant naturally contains an enzyme that rapidly breaks down and detoxifies the herbicide. Yet in the Midwest, corn is often used in rotation with soybeans, which are susceptible to atrazine. Sometimes winter climatic conditions cause residues of atrazine to remain in the soil the following spring. Such residues can dramatically reduce the yield of soybeans planted the following year.

To avoid such losses and to broaden the range of each herbicide, plant breeders are interested in developing herbicide-resistant crops. An atrazine-resistant soybean, for example, would be ideal for the Corn Belt.

But before the trait can be transferred, it must be understood. At

Michigan State University, Charles Arntzen and colleagues have been trying to elucidate the genetic mechanism of atrazine resistance in weeds. Atrazine-resistant weeds began to appear spontaneously in the early 1970s, primarily in areas of prolonged, steady use of the herbicide. Some 30 species of weeds have now become atrazine resistant. To find out how, Arntzen, a plant biochemist, drew on the work of plant physiologists, other biochemists, and agroecologists, as well as molecular biologists.

Like the majority of herbicides, atrazine acts in the chloroplasts where it disrupts photosynthesis. Photosynthesis is the process by which plants convert sunlight to chemical energy. Through photosynthesis, the sun's energy is used to support all higher forms of life on earth.

In photosynthesis, sunlight is absorbed through chlorophyll, the green pigment in the chloroplast. The sun's energy is used to force electrons to an excited state. In a complex series of reactions, the energy held by the excited electrons is used to build carbohydrates.

One of the keys to photosynthesis is the transfer of solar energy from the chlorophyll, where it is absorbed, to the place where it is used. The excited electrons hold that energy. In bucket-brigade fashion, electron carrier molecules transport those electrons from one molecule to another. One of these carrier molecules is quinone.

It is now known that atrazine kills plants by disrupting this electron transport, thereby blocking photosynthesis. It does this by competing with quinone, one of the carrier molecules. When atrazine is taken into the chloroplast, it can take the place of the quinones on one type of protein in the membrane that holds the electron transport chain together. Without quinone to transport the electrons, photosynthesis is halted.

Atrazine resistance arises from a mutation that alters that membrane protein so that it will no longer bind atrazine. The quinone, however, still binds to the altered membrane protein; consequently, electron transport remains undisturbed in the presence of atrazine.

Using traditional biochemical procedures, Arntzen's group spent three years trying to purify that protein in order to study the nature of the mutation. They did not succeed. The task of determining the sequence of amino acids in a protein is difficult and time-consuming. It is easier to sequence the nucleotide bases in the gene that codes for the protein. From other studies, Arntzen knew that the protein was encoded by a gene in the chloroplast. Arntzen's group began collaborating with Lee McIntosh of Michigan State University and Lawrence Bogorad of Harvard to study the genes of the chloroplast. One of the chloroplast genes had already been isolated and cloned. It was the gene Arntzen was seeking.

Since then, the genes in both atrazine-susceptible and atrazine-resistant weeds have been sequenced. The only difference between the two is one nucleotide base: in the resistant weed, an adenine is replaced by a guanine. That one change spells one different amino acid, creating a slightly different protein—but different enough to cause a "glitch" that prevents atrazine binding. In short, one nucleotide substitution in a single gene determines whether a plant is resistant or susceptible to this herbicide.

## Engineering an Atrazine-Resistant Crop

Because resistance to atrazine is conferred by a single gene, this trait seems amenable to molecular gene transfer. But there are other, simpler approaches to develop herbicide-resistant crops.

To date, the greatest success has come from classical plant breeding— aided by knowledge garnered from molecular-level investigations. At the University of Guelph, Ontario, a research team including W. Beversdorf and Vince Souza-Machado have bred atrazine-resistant strains of oilseed rape (*Brassica napus*) and summer turnip rape (*Brassica campestris*). This was accomplished through repeated backcrosses between rape and a closely related resistant weed, wild turnip (*Brassica campestris*). The key to this success was the knowledge that the trait is carried in the chloroplast. In sexual crosses, the chloroplast is transmitted by the maternal line alone; the male contributes only nuclear DNA. That means the only way to generate a resistant crop is to use the resistant weed as the female parent and a crop plant as the source of pollen. The resulting resistant progeny are then fertilized with pollen from the crop plant, and the process is repeated for five to seven generations. "Finally what you get is a crop—a new plant—in which the cytoplasm, including the chloroplast, is essentially donated by the weed, and the nucleus is donated by the crop plant," Arntzen described.

Although backcrossing is laborious, it works. "Current estimates suggest that by 1985, with increased seed stocks, there will be close to one million acres of atrazine-resistant oilseed rape grown in Canada. It's a dramatic success story, and it didn't take one iota of genetic engineering."

Unfortunately, an identical approach is not feasible with many crops, as few are cross-fertile with weeds. There are, however, many herbicide-resistant weeds that are closely related to—but not cross-fertile with— major crop plants. For example, the atrazine-resistant weed black nightshade (*Solanum nigrum*) is in the same genus as potato (*Solanum tuberosum*) and the same family as tobacco and tomato. That is where the

Atrazine-Resistant Weeds That Are in the Same Botanical Family as Crop Plants

| Atrazine-Resistant Weed | Crop Plant |
|---|---|
| CHENOPODIACEAE | |
| Chenopodium album (common lambsquarters) | Beta vulgaris (sugar beet) |
| Artiplex patula | Beta vulgaris (red beet) |
| COMPOSITAE | |
| Senecio vulgaris (common groundsel) | Helianthus annuus (sunflower) |
| Ambrosia artemisifolia (ragweed) | Carthamus tinctorius (safflower) |
| CRUCIFERAE | |
| Brassica campestris (wild turnip) | Brassica campestris (turnip rape) |
| | Brassica napus (oilseed rape) |
| | Brassica oleraceae (cabbage) |
| SOLANACEAE | |
| Solanum nigrum (black nightshade) | Solanum tuberosum (potato) |
| | Lycoperiscon esculentum (tomato) |
| | Nicotiana tabacum (tobacco) |

SOURCE: Charles J. Arntzen, Plant Research Laboratory, Michigan State University, East Lansing.

new genetic technologies come in. Because the species are closely related, their protoplasts can be fused in culture to create a hybrid. The first of these experiments was reported in 1982 by Horst Binding in West Germany and Jonathan Gressel in Israel and their colleagues, who fused protoplasts of black nightshade and potato. The goal was a resistant potato; unfortunately, the atrazine-resistant hybrid was more like the weed than the potato. A solution may be in sight, according to Arntzen. In a half dozen laboratories around the world, researchers are now trying to inactivate the nucleus in the protoplast from the donor weed, so that just the weed's cytoplasm—which contains the resistant chloroplast—will be introduced into the potato protoplast. "If somebody doesn't have a herbicide-resistant potato plant within the next year or two, I'd be very surprised," Arntzen said.

An alternative approach is to find a mutant in nature—yet this particular mutation rarely occurs. Arntzen and others are investigating methods to induce this mutation.

The most powerful technique, if it can be mastered, will be to transfer the resistant gene from a weed into a crop plant using recombinant DNA technology. "There has been a lot of progress along these lines," Arntzen said, "but we still have a great deal of work left to do." So far, the gene for herbicide resistance has been isolated and cloned inside a bacterium. Now the Michigan State and Harvard University groups are trying to achieve gene expression. The biggest hurdle will be finding a vector to carry the herbicide resistance into a crop, since the gene must be inserted into the chloroplast. The only successful plant vector that has been developed to date—the Ti plasmid—does not work for chloroplasts.

## Bioengineered Microorganisms to Combat Plant Diseases

One of the most promising, and relatively unexplored, applications of the new genetic technologies is in combatting soil-borne plant diseases, according to Milton N. Schroth, a plant pathologist at the University of California at Berkeley. Some of the most destructive plant diseases, such as *Fusarium* wilts and *Phytophthora* root rots, are caused by microorganisms that inhabit the soil. Other, less virulent microorganisms also exact their toll: the presence in the soil of low-grade disease agents can lower yield significantly. For example, if strawberries are grown on fumigated soil, in which all the soil microorganisms have been destroyed, the yield is approximately 20 tons per acre, or four times higher than if they are grown on unfumigated soil. Because the effects of soil pathogens are sometimes insidious, it is difficult to estimate exactly the economic costs of soil-borne diseases, but they undoubtedly can be considerable.

The increasing adoption of minimum or no-till farming practices provides an extra incentive for developing effective controls. Both of these practices leave organic debris on the soil surface, which makes the soil both wetter and colder—creating a more favorable environment for pathogenic microorganisms.

There are several possible strategies. Fumigation of the soil effectively controls disease, but it is costly and impractical for large areas. Moreover, its benefits are transitory, as disease-causing microorganisms can be easily reintroduced by wind or animals.

Another approach is to breed resistant cultivars. Many resistant varieties have been developed, yet for unknown reasons the introduction of disease resistance often results in a loss in yield. Tobacco plants bred for resistance to *Fusarium* wilt fungus, for example, commonly show a nearly 6 percent reduction in yield as compared to susceptible plants.

Resistant plants could also be developed through gene-splicing techniques. "It would be ideal if genes conferring resistance to pests could be introduced to the plant, expressed, remain stable, and not result in a cost to the plant," Schroth said. Unfortunately, demonstrated techniques for doing this are not yet available. Moreover, it is not clear

Reductions in Yield and Quality in Disease-Resistant Tobacco Lines in Comparison to the Susceptible Line[a]

| Resistance | Percent Yield Reduction/ha | Percent Price Reduction |
|---|---|---|
| Tobacco mosaic | 5.9 | 1.5 |
| *Fusarium* wilt | 6.9 | 1.9 |
| Mosaic + *Fusarium* wilt | 9.9 | 1.9 |
| Mosaic + bacterial wilt | 7.1 | 2.7 |
| Mosaic + root knot nematode | 5.0 | 1.8 |
| *Fusarium* wilt + bacterial wilt | 10.3 | 4.0 |
| *Fusarium* wilt + black shank | 6.6 | 3.2 |

[a] Based on data given in Chaplin, 1970. Agron. J. 62:87–91.
SOURCE: Milton N. Schroth, Department of Plant Pathology, University of California at Berkeley

whether the introduction of resistance through gene-splicing would result in the same loss in yield as does conventional plant breeding.

The alternative genetic engineering approach is to harness and improve upon the beneficial microorganisms that inhabit some soils and use them to combat plant disease. In nature some soils, known as disease-suppressive soils, contain beneficial microorganisms that help to protect plant roots from pathogens. The mechanism of disease suppression in these soils is poorly understood, and it does not come from the beneficial microorganisms alone. Instead, it seems to be controlled by complex interractions among both biotic and abiotic factors. It has long been known, for instance, that physical conditions such as the salinity, acidity, temperature, and moisture levels of soils can render plants less or more susceptible to a disease. Nonetheless, beneficial microorganisms seem to play a major role. If a disease-suppressive soil is fumigated to destroy all microorganisms, the soil loses its capacity to suppress plant pathogens. But when some of the unfumigated suppressive soil is reintroduced, the disease-suppressive quality of the soil is restored.

These naturally suppressive soils provide a substantial boost to crop yield. In Provence in southeastern France, the suppressiveness of the soil varies greatly from region to region. This has been determined by infesting the soil with pathogens and then comparing the severity of the ensuing disease. For centuries, muskmelons have been grown in the Chateaurenard area with little trouble from *Fusarium* wilt, even though the fungus is present in the soil. Yet in the neighboring two regions of Cavaillon and Carpentras, the disease can be so severe that the muskmelon crop sometimes has to be abandoned.

Little is known about the microorganisms that inhabit the rhizosphere, or soil-root zone. Most research to date has focused on nitrogen-fixing bacteria because of their potential to reduce the need for chemical fertilizer (*see* Nitrogen Fixation, p. 48). In just a few years, molecular biologists have made great strides in understanding the genetic control of this trait. The same tools can be used for studying, and ultimately improving, these other beneficial microorganisms.

Some of the most promising candidates for biocontrol agents are the root-colonizing bacteria, generically known as rhizobacteria. Some of these have the beneficial effect of promoting plant growth; others have deleterious or neutral effects. For use as biocontrol agents, the bacteria must be able to colonize the roots aggressively and have the potential to dominate the ecological niche. Finding such bacteria will be difficult,

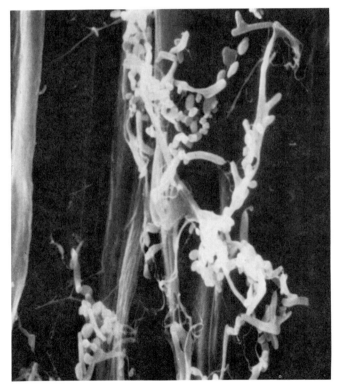

*Pseudomonas* colonizing the surface of a sugar beet root. This scanning electron micrograph shows chainlike colonies of bacteria against the ribbed background of a sugar beet root ($\times$ 3000). These beneficial bacteria suppress the growth of plant pathogens that could otherwise attack sugar beets. Courtesy of Milton N. Schroth, Department of Plant Pathology, University of California at Berkeley.

as less than 5 percent of bacteria isolated to date from plant roots are able to colonize the roots effectively and promote plant growth. Several strains of *Pseudomonas* can.

Rhizobacteria seem to work through two generic mechanisms. One is antagonism—the bacteria compete with and displace the deleterious organisms on the plant root. The second is inhibition—some rhizobacteria produce antibiotics that inhibit a variety of pathogens.

It seems likely that disease protection in soils is conferred through a variety of microorganisms and favorable environmental conditions. The hope is to identify some of the key beneficial microorganisms and adapt them to use as biocontrol agents in conducive soils, either by manipulating the soil environment or modifying these microorganisms to improve their efficiency. That, of course, depends on an understanding of their normal mode of action.

Research to date has been promising. Though relatively little is known about rhizobacteria, their application to seeds and roots at planting time can increase plant growth and yield. In Idaho, California, and Pennsylvania, potato yields increased 5 to 33 percent following application of *Pseudomonas*. It has also been effective on sugar beets and radishes.

It might be simpler to introduce the microorganisms directly into the soil, but that approach is "unreachable" at this time and may remain impractical for commercial agriculture, Schroth said. The biocontrol agents would have to compete with the other organisms already present in the soil. Those long-term residents are "well entrenched, and not easily displaced by intruders." In such a scheme, vast amounts of inoculum would be required on a regular basis; the cost might be prohibitive.

Because strains of the same species are best adapted to occupying the same ecological niche, the ultimate approach may be to engineer the pathogen to control itself. This would entail inactivating the disease-causing agent from the microorganisms and then using this disarmed bug to displace its pathogenic relative. Eventually, when more is known about complex interactions in the rhizosphere, it may even be possible to manipulate the soil ecosystem to favor beneficial microorganisms.

Identifying and improving rhizobacteria will require the combined efforts of bacterial ecologists, plant pathologists, biochemists, and genetic engineers. Specifically, they need to determine the genetic factors that govern root colonization. They need to identify the key factors that enable a microorganism to compete successfully in an ecological niche. One means of increasing their competitive ability might be to bioengineer them to tolerate greater moisture stress.

Schroth cautioned against underestimating the complexity of the agroecosystem. It may be possible to design a strain of bacteria that gives dramatic results in the laboratory. But in the field, competing with microorganisms that have evolved for hundreds of thousands of years, a successful laboratory strain might not perform well, or even survive.

## Nitrogen Fixation

Nitrogen is an essential plant nutrient and a key determinant of crop productivity. Unfortunately, the nitrogen content of agricultural soils is quickly depleted. Farmers worldwide supplement the available nitrogen with some 60 million metric tons of nitrogen fertilizer annually. By the year 2000, an estimated 160 million metric tons of nitrogen fertilizer may be used each year. Producing that fertilizer is both expensive and energy intensive.

As farmers face the prospect of rising bills for nitrogen fertilizer, their crops are literally being bathed in nitrogen gas, as roughly 80 percent of the air is nitrogen. Yet plants are unable to use nitrogen from the air. Soybeans and other legumes are an exception; they have a symbiotic relationship with nitrogen-fixing bacteria, *Rhizobium*. In some soils, where *Rhizobium* are indigenous, the farmer need only plant the legume. In areas where the bacteria are not present, the farmer adds or inoculates the soil with *Rhizobium*. In either case, no nitrogen fertilizer is necessary. The *Rhizobium* infect the roots of the plants, causing nodules to form. Inside the nodules, millions of bacteria convert the nitrogen that is in the air to ammonia, which the legume, like other plants, needs for protein synthesis.

Agricultural yields could be sustained at tremendous savings if biological nitrogen fixation can be improved and extended to major crops, such as corn and wheat, that now depend on costly nitrogen fertilizer.

The cluster of genes that control nitrogen fixation in microorganisms has been isolated and analyzed. In numerous academic and industrial laboratories, researchers are trying to understand how those genes are regulated and how they can be used in practical crop improvement schemes. Winston Brill of the University of Wisconsin and Cetus Madison, Corp., described that work.

The earliest payoff may come from attempts to improve the efficiency of nitrogen fixation in legumes. The approach is to engineer either the plant or the *Rhizobium*, or both, to improve the symbiotic relationship. According to Brill, genetic manipulation of the bacterium is far simpler than manipulation of the plant. He has used both standard mutagenesis and recombinant DNA techniques to develop improved strains of *Rhizobium*. When inoculated with the mutant strains, plants show increased vigor and growth.

 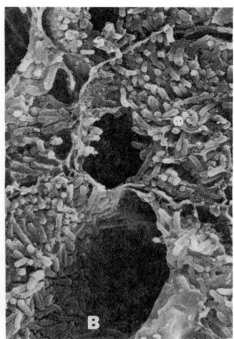

Root nodules in legumes. The nitrogen-fixing bacteria, *Rhizobium*, can induce nodules to form on the roots of legumes (A). The nodule cells become filled with bacteria (B). In this symbiotic relationship the *Rhizobium* are dependent on a carbohydrate energy supply from the plant and, in turn, the bacteria reduce atmospheric nitrogen to a chemical form that the plant can utilize.

Most effort, however, is concentrated on extending nitrogen-fixing abilities to other crops. Again, there are several approaches. Brill has been working with another nitrogen-fixing bacteria, *Azotobacter*, trying to engineer it to supply nitrogen for corn. Unlike *Rhizobium*, *Azotobacter* do not infect and form nodules on the plant roots; instead, they are free-living in the soil. The *Azotobacter* leak part of the nitrogen they fix into the soil. In turn, they are dependent on a supply of energy-rich carbohydrates that are exuded in part from plant roots. Brill's experiments involve manipulating genes of both the plant and the bacteria— for example, inserting genes into the bacteria to enable them to adhere tightly to corn roots for a more efficient passive association. The technique has had some, albeit modest, success. In the laboratories, corn plants inoculated with *Azotobacter* have obtained up to 1 percent of their total nitrogen from this process. "That's not much," Brill conceded, "but it's a start."

The most efficient means of supplying nitrogen would be to transfer the nitrogen-fixing genes from the bacteria into the plant. It is also one of the toughest tasks imaginable. A number of laboratories have isolated the nitrogen-fixing genes from *Klebsiella*, bacteria similar to *Rhizobium* but easier to work with in the laboratory. Nitrogen fixation is a complex, multigene trait controlled by a cluster of 17 genes. These genes are broken down into smaller units, each of which is regulated separately. To endow a plant with the ability to fix its own nitrogen would mean transferring all 17 genes, along with the complete collection of regulatory signals.

## PROBING THE MECHANISMS OF NITROGEN FIXATION

Soybeans, alfalfa, and other legumes have a symbiotic relationship with the bacteria *Rhizobium* that enables these plants to obtain nitrogen from the soil. The increasing interest in extending the ability to fix nitrogen to other crops has spurred efforts to understand the unusual relationship between *Rhizobium* and legumes.

When *Rhizobium* infect plant roots they cause the cells to proliferate, giving rise to nodules on the roots. *Rhizobium* induce another, apparently unique, change in the plant cells: at the spot where the bacteria first come in contact with the root, the plant cells form a tubularlike structure, known as an infection thread. These infection threads wind throughout the cells in the root nodule, providing a conduit through which bacteria migrate from one cell to another. Once inside the cells, the bacteria convert nitrogen to a chemical form the plant can use.

The top photograph, a scanning electron micrograph (magnification × 2000), shows an infection thread traversing a cell in an alfalfa root nodule. At the end of the thread (upper left) bacteria are being released into the cell.

Bacteria other than *Rhizobium* are not known to induce the formation of infection threads. Research is under way, using gene-splicing techniques, to transfer the genes that control nodule formation (the nod genes) and nitrogen fixation (the nif genes) from *Rhizobium* to other bacteria. If bacteria that infect nonleguminous plants can be endowed with the properties of *Rhizobium*, it might offer a means of extending nitrogen fixation to other crops.

The nod and nif genes have recently been transferred from *Rhizobium* to *Agrobacterium tumefaciens*. When the genetically engineered *Agrobacterium* infected an alfalfa plant, it induced root nodules and infection threads to form (bottom photograph, magnification × 760). The genes for nitrogen fixation, however, were not expressed.

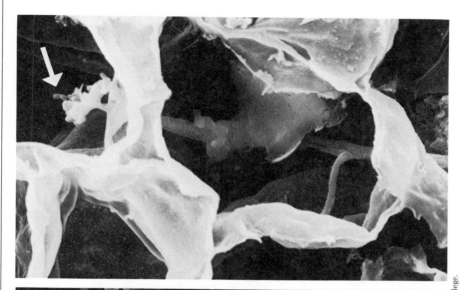

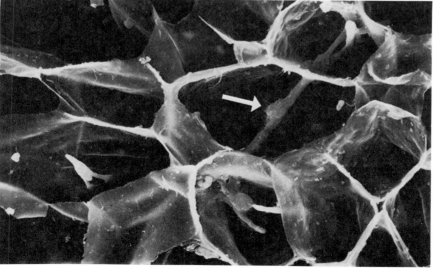

Courtesy of Ann W. Hirsch, Department of Biological Sciences, Wellesley College.

Brill predicted that this cluster of genes will soon be moved into dicots such as potatoes and tomatoes and later into monocots like corn and other cereals. "Is that the end of the story? Do we now have nitrogen-fixing corn?" he asked. The answer is no. The entire cluster of genes has already been transferred into yeast, but the genes were not expressed.

Not only must gene expression be achieved, but the genes must be inserted into the proper place in corn and turned on at the appropriate time. Nitrogen fixation requires an energy-rich, oxygen-depleted microenvironment—the corn's mitochondria or chloroplasts might be a suitable site for the genes. In addition, the chemical reactions involved in nitrogen fixation require far more iron and molybdenum than are normally found in most plant cells. It is too early to tell if the hurdles are simply formidable or if they are insurmountable.

## The Bottom Line

The final test of these new agricultural products—the improved varieties and bioengineered microorganisms, for instance—will be their performance in the marketplace. The new products must offer an advantage over existing ones if the farmer is to adopt them. As Schroth explained, the bioengineered products must improve the profit margin per hectare—either by increasing yield or reducing production costs. That, in turn, depends on how well the new products can be integrated into existing agricultural techniques.

Though disease-resistant plants can be bred, they often have reduced yield. It may be more economical for the farmer to use a susceptible variety and a fungicide to control the disease. Similarly, the benefit of a biocontrol agent will have to be weighed against a fungicide. Key factors influencing the commercial success of a biocontrol agent might be its shelf life and the ease with which it can be applied to the soil.

So far, many researchers have been so caught up in what is scientifically possible that they have neglected the practical considerations, such as market analysis, Schroth added. He suggested that molecular biologists work with plant breeders, agronomists, and pomologists in identifying scientifically and economically attractive projects for genetic engineering. "It will not be a simple task to improve productivity per hectare," he said. "And it certainly will not be done by the unilateral efforts of one discipline."

Bogorad concurred. "It is clear that we need molecular biologists plus plant pathologists and agronomists—people who know about real plant problems. One of the difficulties we have today is that there are very, very few people who understand both sides of the problem."

# Policy and Institutional Considerations

> Our technical and personnel resources are still painfully limited for a major push into the genetic engineering of plants. We need substantial improvement in experimental techniques and more knowledge about plant systems. We need modern equipment and instrumentation to take advantage of experimental opportunities now, and we're going to have to develop new kinds of laboratory tools for the future.
>
> But most important, we need people. Too few—far too few—students have chosen careers in plant sciences. And that's primarily our fault, not theirs. By our federal funding patterns, we've made it too attractive for them to apply their talents in other fields of biology and science.
>
> George A. Keyworth
> *Science Adviser to the President*

The application of genetic engineering to agricultural problems is not simply a matter of developing appropriate vectors and cell culture techniques. Before molecular biologists can modify and improve plants with any predictability, they must understand the physiology, biochemistry, and genetics of plants. It is no accident that the first successes of genetic engineering have taken place with bacteria such as *Escherichia coli*, which have been studied intensively for years.

The understanding of the detailed biology of plants lags far behind that of microorganisms and some animals. Fundamental questions remain about plant development, growth, and metabolism, as well as the biology and ecology of host-pest relationships. The effects of environmental stresses need to be studied. Similarly, additional research is needed on the susceptibility and resistance to disease, as well as the cellular, molecular, and genetic basis of pathogenicity. Molecular biol-

53

ogists must be able to identify the agriculturally important genes, as well as elucidate gene expression and regulation.

The problem, as George A. Keyworth pointed out, is that there are too few scientists studying the basic biology of plants. There are fewer still who are equipped to translate fundamental knowledge into effective genetic techniques for crop improvement.

Funding for basic plant research, especially for plant molecular biology, has been scarce. Most of the support for plant science comes from the National Science Foundation (NSF) and the U.S. Department of Agriculture (USDA), though other agencies, such as the U.S. Department of Energy (DOE) and the National Institutes of Health (NIH), also support several areas of basic plant research. The research funded by the NSF and NIH programs is primarily investigator initiated. Scientists submit research proposals on subjects of their own choosing, the proposals are evaluated for scientific merit by a peer group of scientists, and the funding is awarded on a competitive basis. At NSF, approximately $41 million was spent in fiscal year 1982 on investigator-initiated grants in plant science, primarily through the Directorate for Biological, Behavioral, and Social Sciences. That total includes all plant science ranging from ecological and biosystematics studies to research at the molecular level. Opportunities for predoctoral and postdoctoral training fellowships in plant sciences have also been limited. From 1974 to 1979, the federal government supported 23,420 postdoctoral fellows in the biological sciences, 15,845 in the health sciences, and only 821 in plant science. It is not surprising, Mary Clutter of NSF said, that the best students have chosen careers in other fields of biology where they have a better chance of obtaining adequate support—both during training and later during their academic research careers.

USDA's Agricultural Research Service (ARS) has a sizeable in-house research effort. In 1982, ARS spent about $165 million for research on plants. Research in plant biotechnology, including research on photosynthesis, biological nitrogen fixation, cell culture, molecular genetics, plant stress, and plant growth regulators, represented approximately $7 million of that total. The majority of the research funded by ARS is targeted to specific problems. Through the Hatch Act, USDA also provides funds to land grant universities for partial support of the agricultural research programs in the 50 states and territories. Yet, unlike NSF or NIH, USDA does not have a strong program to support investigator-initiated research. The exception is the small USDA Competitive Grants Program. In fiscal year 1982, $16.3 million was available for competitive grants; of that, only about one-fifth of the grants funded were in the field of plant molecular biology.

## Neglect of Basic Science

Historically, agricultural research in the United States has been oriented toward applied questions both in the ARS and in the land grant universities, which are supported by a combination of federal and state funds. In the states, this emphasis is in part a result of the mandate of the land grant system, which was created in 1862 to perform scientific research of direct benefit to agriculture. Over the years, the land grant universities have developed a variety of practical innovations, ranging from improved seed and horticultural practices to better farm machinery and irrigation systems. "All of these developments involved serious and sometimes brilliant science," said Lowell N. Lewis of the California Agricultural Experiment Station. "But overall the effect of this emphasis—whether real or perceived—has all too often been to relegate science for agriculture within the land grant system to the status of second-class research, narrowly conceived to meet the need of farmers for technical improvements. This system has worked remarkably well in fostering agricultural productivity, but in the kinds of problems we confront in agriculture today, it appears to have been something of a mixed blessing. The same system that by and large has excelled at sustaining effective relations with the agricultural community has, over the last few decades, contributed to the declining association of agricultural research with basic science."

One of the costs, Lewis said, is that in the last four years enrollment in agricultural colleges has declined by 15 percent—"at least partly because of the students' perception that the curricula in agriculture-related science do not represent the intellectually and professionally exciting studies on the cutting edge of science." Even now, according to the National Association of State Universities and Land Grant Colleges, 13 percent of jobs in agriculture go unfilled for lack of qualified college graduates in the agricultural sciences. Lewis predicted that this percentage will continue to rise.

Similar problems plague the federal agricultural system, said George E. Brown, Jr., a U.S. congressman from California and chairman of the House Agriculture Subcommittee on Department Operations, Research, and Foreign Agriculture. "Over the past 10 years there has been a decline in support for the agricultural research establishment and a failure to maintain its position on the frontiers of knowledge. A freeze on personnel has kept capable young researchers from entering the federal system, and many of them have been forced to seek opportunities elsewhere. Plant biology has suffered from a lack of new blood and neglect of basic work. We have emphasized applying existing knowledge and failed to replenish our intellectual capital."

To some extent, Congress shares the blame for this situation, Brown conceded. "In too many cases, we have meddled with the system to ensure direct, short-term benefits to particular areas. We have protected established programs rather than pressing for new ventures."

The solution, Brown and others agreed, is stable, long-term funding for basic research. A quick infusion of funds, narrowly targeted to a specific problem, will not suffice.

Both the ARS and the land grant universities must strengthen their programs in fundamental science, Lewis said. In addition, the plant research community should recognize and draw on the traditional strengths of the state agricultural experiment stations. These stations provide a breadth of expertise in applied agricultural sciences, valuable field research facilities, and a tradition of close communication with the agricultural community. The experiment stations can help plant genetic engineers in developing practical new strategies for crop improvement. Opportunities for such active collaboration should be developed, Lewis suggested, adding that the concept of "agricultural science" should be expanded to include the total scientific community.

Ronald Phillips, a plant geneticist at the University of Minnesota, reminded the participants that basic research pertinent to biotechnology is performed not only at federal laboratories and land grant universities but also at private universities, research institutes, and in industry. He and other speakers asked whether the USDA was providing the leadership needed in developing the basic sciences necessary for improved agricultural technology. Phillips suggested establishing an oversight group, composed of leading scientists from both public and private research programs, to identify promising opportunities for research and training in basic science related to agriculture.

Today there are signs that the climate for research in plant biology is beginning to change. "We cannot overemphasize the importance of developing our nation's human capital," Orville Bentley, assistant secretary of USDA said, recommending a two-part approach to strengthen university programs. One would be a competitive graduate fellowship program "targeted at those areas with the most critical shortages of expertise." The other would be grants to universities to enhance educational programs in plant science through faculty training and improved instructional equipment, for instance.

As Keyworth pointed out, the fiscal year 1984 budget includes a 20 percent increase in funds for basic biology at NSF with an emphasis on support for the plant sciences. The budget also proposes funding for new postdoctoral research positions in agriculture at USDA. Keyworth acknowledged that these federal efforts are only "fragile thrusts."

At NSF, the Division of Physiology, Cellular and Molecular Biology this year started a postdoctoral fellowship program for the plant sciences. It received 196 applications but had funding to make only 24 awards, according to section head Mary Clutter. The division is also attempting to increase the grant budgets for top plant biologists to enable them to support graduate students and postdoctoral fellows. Clutter also pointed out that the nonprofit McKnight Foundation recently awarded individual research grants and six training grants in basic plant sciences that will amount to about $5 million over the next three years. "What the federal government has been unable to do," she said, "a private foundation has decided to do in a small but significant way."

Industry is also a major supplier of research, according to Ralph W. F. Hardy, director of life sciences research at E. I. du Pont de Nemours & Co., Inc. He said that in 1981 in the United States expenditures for agricultural research and development, broadly defined to include everything from food and forest products to farm equipment to fertilizers, were estimated in excess of $5 billion. Expenditures by the private sector accounted for roughly 60 percent.

Hardy sees the roles of the public and private sectors in agricultural research as distinct but interdependent. To the private sector properly falls the responsibility for advancing basic knowledge and training future scientists. Industry draws on that resource in developing competitive agricultural products. In addition, as knowledge increases about genetic engineering, it will provide a more realistic framework for regulation.

By contrast, the private sector selects parts of that system where it thinks it can make a contribution and get a return on its investment. "The private sector's job is to discover, develop, manufacture, and market proprietary products, services, and processes," Hardy said. "It will, in doing this, generate some new knowledge. It will also provide some support for the public sector. But it should in no way be looked to as a major supporter of public sector activities."

## Multidisciplinary Training

Not only will plant biotechnology require more scientists, it will require scientists having different, broader training. "The most rapid gains in applying this technology to plant and animal science will come when applied and basic sciences collaborate in common research programs," said Charles Hess, dean of the College of Agriculture and Environmental Science at University of California at Davis. "For example, the combination of molecular biologists with plant breeders and plant pathologists will accelerate the genetic engineering of disease-resistant plants." Hess

Courtesy of U.S. Agency for International Development.

suggested that USDA provide additional funds through its Competitive Grants Program for such collaborative biotechnology research.

Whether working in the laboratory or the field, scientists with backgrounds in agronomy and molecular biology will need to be able to communicate. Thus, a crucial component of the training of both future agricultural scientists and molecular biologists will be a grounding in the related disciplines.

Such training has been sorely lacking to date. According to Philip Filner of the ARCO Plant Cell Research Institute, "In my experience recruiting and interviewing many young scientists in the last two years, I have the feeling that they have struggled very hard to master the jargon of their field of expertise and have then become addicted to it. They are not very good at communicating their ideas. There is a need to improve their ability to understand, appreciate, and use the main ideas of complementing fields. I've encountered graduate students who are working on the structure of mitochondrial DNA who question the necessity of understanding the details of energy metabolism in the mitochondria. Similarly, Ph.D.s working on a bacterial virus may be almost totally ignorant of animal or plant viruses. They simply don't understand or appreciate what others do and how they do it. This is a very serious disadvantage when they come into an industrial environment where the emphasis is on collaboration toward a common goal."

Filner suggested that students of plant molecular biology also learn plant physiology and breeding. Kenneth J. Frey, an agronomist at Iowa State University, added that plant breeders must gain a familiarity with molecular biology.

In some cases, it may be sufficient to simply supplement major studies in one field with coursework in another. In other cases, more extensive multidisciplinary training may be necessary. A number of scientists trained in the molecular biology of microbial or animal systems are now being drawn to plant genetic engineering. An opportunity exists to attract more students to plant research through special workshops in plant biology and postdoctoral research positions to work on plant science and agricultural problems.

# University-Industry
# Relations

The commercialization of genetic engineering has created new ties—and new tensions—between industry and the university. Major corporations were caught largely unprepared by the rapid advances in biotechnology, having neither the expertise nor staff to enter the field. To do so, they have drawn heavily on the resources of the university, either through establishing consulting or contracting agreements with individual researchers or research partnerships with the university.

By contrast, academic researchers were among the first to realize the vast economic potential of the new genetic technologies. In the mid-1970s, some university scientists launched their own biotechnology firms where they could translate their ideas into products and profits (*see* Starting a Biotechnology Company, p. 65). These firms were often located a stone's throw from the university, enabling the researchers to divide their time between their industrial and academic laboratories. Some scientists left the university altogether. Others remained in the university but assumed managerial or equity positions in the new biotechnology firms. Originally, there were fears that the best scientists would be lured away from the university by the promise of vast profits. Although such fears have lessened, universities now face strong competition in filling faculty openings in these areas.

These new research partnerships offer advantages to both parties. For universities, they offer funds for research and instrumentation in a time of slack federal support. In return for their investment, industry gains access to scientific expertise, a training ground for industrial scientists and, usually, rights to any patentable discoveries. Society, too, gains, for such collaboration can accelerate the pace of technological innovation.

Nonetheless, many in the academic community are concerned that this new infusion of corporate dollars will divert the university from its primary mission of fundamental research and teaching. Yet, as Gilbert Omenn, dean of the School of Public Health and Community Medicine at the University of Washington pointed out, corporate funding of university research is not new. Indeed, industry was the largest source of outside funds for the university before the federal government assumed that role following World War II. Some institutions, such as Stanford University and the Massachusetts Institute of Technology, have a long history of close ties to industry. At these and numerous other universities, faculty members have long consulted for or obtained research grants from industry.

These new relations, however, differ in some key ways. For one, most of the previous arrangements involved faculty in engineering and business—not in basic biology. Moreover, the sheer size of some of these new agreements sets them apart from those that preceded them. Hoechst AG, a West German chemical firm, is spending $70 million over the next 10 years to develop a department of molecular biology at Massachusetts General Hospital, a Harvard University affiliate. Washington University has a 5-year, $23.5 million contract with Monsanto Co. for research on peptides and proteins.

## University Concerns

The central fear is that the open communication among professors and between professors and students will be stifled by the need to protect proprietary information. Some wonder if graduate students whose professors are supported by corporate funds will be forbidden to present their research results at seminars or conferences. Others wonder if there will be a subtle reorientation of values when an entire department is supported by a corporation.

Patent rights pose yet another concern. In the past, universities have profited from patent and license arrangements—the University of Wisconsin Alumni Research Foundation is one example. However, the granting of exclusive rights to one company—which seems to be requisite in some new university-industry partnerships—conflicts with university tradition of open dissemination of information. According to Richard S. Caldecott, dean of biological sciences at the University of Minnesota, universities will confront a number of choices and will undoubtedly come to different solutions. They must decide who will hold the patent, the university or the industrial sponsor, as well as how royalties will be assigned—to the department, the university, the investigator, or all three. They must decide when and if to grant an exclusive license to an industrial sponsor. To assure its claim to exclusive rights, an industrial sponsor may insist that its funds be kept separate from any other support for the laboratory. There is concern that federal and private funds may sometimes be comingled.

To Omenn, the "people problems" are of the greatest concern—"the delicate, crucial, and rather complicated relationships between the individual faculty member and the university—the difficult matter of avoiding conflict of interest and protecting the intellectual property of all colleagues involved in an academic enterprise."

Another concern is that scientists who hold equity or managerial positions in a commercial venture may neglect their students to perform more lucrative projects for their industrial sponsor. When a professor has two employers, graduate students and postdoctoral fellows are particularly vulnerable. For instance, when profits are involved, the problem of ensuring that students receive proper credit for their work may be further aggravated.

"The only way to deal with the potential conflicts of interest is openness—full disclosure," Omenn said. "The students and faculty members should know the kinds of external relationships other faculty members are engaged in."

Already, several universities have adopted conflict-of-interest policies that require full disclosure of faculty arrangements with outside corporations. Some universities prohibit full-time faculty members from holding equity or managerial positions in a commercial firm.

Many of the potential conflicts can be avoided if terms are thoroughly and clearly defined in the contract. For example, many of the recent agreements require that the researchers submit an article to the corporate sponsor for review prior to submission to a journal. Cornell University has taken a slightly different tack. According to Theodore Hullar, the university's director of research, in an effort to safeguard open communication and protect the rights of students, Cornell contracts specify that if a graduate student is involved in corporate-sponsored research, those results can be discussed in any seminar on campus.

## Industry Concerns

In the debate over university-industry relations, said Reuven M. Sacher, director of biological research at Monsanto Co., it should not be forgotten that industry also has an interest in preserving academic values. The strength of the university stems from its freedom to pursue ideas wherever they lead, without pressure to meet practical goals. Moreover, a strong research effort is an essential part of training of future scientists and engineers on which industry depends. To sacrifice the quality of research to meet short-term goals would be killing the goose that lays the golden egg.

Joint university-industry ventures do involve some risk, largely to the universities, Sacher conceded. Yet, that risk is manageable—and outweighed by the benefits of such collaboration to the university, industry, and society. As economist Vernon Ruttan stated, "a rapidly expanding

---

## MAINTAINING THE COMPETITIVE EDGE

The following comments are excerpted from the presentation of Reuvan Sacher, director of biological research at Monsanto Co.

"How can we insure that American industry remains commercially and technologically competitive in the field of biotechnology? Looking at our own company, Monsanto has approximately $7 billion worth in sales in 120 countries, and we spend more than $300 million a year on research and development on nearly 3,500 scientists and engineers in our corporation. Surely Monsanto should be able to compete against Eli Lilly or du Pont in developing biotechnology products. I believe we can. But Monsanto's competition is not merely the du Ponts and the Eli Lillys. Monsanto also has to compete with entire countries in the field of biotechnology.

"The Japanese government has declared biotechnology a national scientific and commercial goal. Japan's Ministry of Trade and Industry has established a consortium of 14 major Japanese companies that will cooperate with the government and universitites in developing biotechnology. The government has set aside $100 million for that purpose. In addition, in 1982 Japanese companies spent more than $200 million apiece in the biotechnology area.

"In the United States, antitrust laws prevent Monsanto, Eli Lilly, and du Pont from conducting research together in most cases. Thus, while Japanese companies are encouraged to cooperate, U.S. companies are enjoined from doing so. Since we cannot cooperate with our 'competitors,' who can we cooperate with? The answer is simple: with American universities. We feel that the research talent in American universities is immense. If you couple that with the development skills of American industry in general, I think we can keep the United States on the leading edge in biotechnology. I think this cooperation can give rise to jobs, useful products, and new ways to meet basic human needs throughout the world."

private sector role in biotechnology research and development is essential if any significant impact of biotechnology on crop yields is to be achieved over the next 20 years." Close collaboration between the university and industry facilitates the transfer of ideas into useful products. Sacher emphasized that collaboration may also be the key if U.S. industry is to maintain its competitive edge.

While Sacher advocated increasing industry support for university research, he cautioned, as did Hardy and others, that corporate funds cannot make up for any shortfall in federal support for university research. In 1981 industry provided 4 percent (about $250 million) of the $6.6 billion spent on university research, Sacher said, adding that it is "inconceivable" that industry support will exceed 6 or 7 percent. The rest must come from government.

### Three-Way Collaboration

Recognizing the economic benefits that can accrue from a strong biotechnology industry, several state governments have launched new efforts to increase collaboration between the universities and industries. Hullar, who sees great promise in these three-way partnerships, described one such effort in New York state.

Through legislative mandate, the state established a public-benefit corporation, the New York State Science and Technology Foundation. Its board of directors includes the presidents or research vice-presidents of the major corporations headquartered in the state—including Eastman Kodak, General Electric, IBM, Xerox, and Corning Glass Works—as well as state leaders in public policy and finance. As one of its functions, the foundation identified eight promising technologies and established a collaborative research center for each.

After a competition with other universities and institutions throughout the state, Cornell University was designated by the New York foundation as its Center for Biotechnology in Agriculture. The uni versity had to demonstrate that it already had excellence in the field and that it could raise industry funds to match the state contributions. Other requirements were that the research have industrial relevance and that the university have an outreach program to the state.

At the same time, Cornell established its own partnership with industry in the state, called the Cornell Biotechnology Institute. Four corporations have already joined, and each will contribute $2.5 million over six years. The institute provides an opportunity for industry scientists to take up residence in a university laboratory for a year, performing fundamental studies not directly related to their corporate pursuits. All research at the institute is basic and nonproprietary, and the university's standard patent policies apply.

---

## STARTING A BIOTECHNOLOGY COMPANY

Within a few years of the pioneering gene-splicing experiments, some 200 biotechnology firms were launched to exploit this new commercial potential. To date, the field has attracted nearly $1 billion in investments. These firms range from small entrepreneural ventures to larger companies such as Genentech, Cetus, Biogen, and Genex. Some well-established corporations have developed their own internal genetic engineering laboratories, while others have invested in the smaller venture capital firms or contracted with them for research.

As an example of some of the factors to be faced in starting a biotechnology company, Anthony Faras spoke at the convocation about Molecular Genetics, Inc., a firm he founded with Frank Pass in 1979 in Minneapolis, Minnesota. While the individual details of each biotechnology venture are unique, their futures all depend on a business plan that includes capital, research personnel and facilities, and potential products to generate revenues. Faras described the first four years at Molecular Genetic, Inc., or MGI—specifically the scientific and economic considerations that have guided the company's development.

### A Business Plan

Both Faras and Pass were professors at the University of Minnesota when they decided to start MGI. Shortly after founding the company, Pass left his position in the dermatology department at the university to become president of the new firm. Faras kept his position as a microbiology professor while simultaneously serving as cochairman at MGI.

In 1979, MGI's first priority was to raise enough money to build a laboratory and hire a staff. That depended on developing a business plan to attract venture capital. Faras and Pass decided to begin by developing veterinary products—specifically vaccines, monoclonal antibodies, and antitoxins to fight infectious diseases, as well as some growth hormones. They planned to expand later into plant biotechnology.

They started with veterinary products for two reasons, Faras said. First, in view of stiff competition from other biotechnology firms, they wanted to get a product on the market quickly. Since they both had expertise in infectious diseases, they realized they could progress more quickly with vaccines than they could in crop improvement. They also avoided human health care products, in large part because of anticipated regulatory constraints.

The second reason was market potential. "Though there are a number of different antibiotics and other types of biologicals used widely, viral and bacterial infections still cause a high proportion of mortality, both in swine and cattle. Equally important, in terms of economic losses, are the morbidity problems." If they could develop an improved product, substantial sales seemed certain.

There is also a vast potential market for genetically engineered crops, though the development time is far longer than that required for vaccines or pharmaceuticals. Once MGI was established in the veterinary area, Faras and Pass reasoned, they would begin work on plants, specifically on improving the protein content and feed quality of corn.

In addition, Faras said, the agricultural focus of MGI fitted well with the community—Minnesota is a large agricultural producer. That might help in raising capital. Perhaps more important, they knew there was a good source of local talent at the university's veterinary and agricultural schools.

### Initial Investors

Business plan in hand, they attracted four investors. "What we got was $1.2 million to develop our first facility and to hire our initial group of scientists. What the venture capitalists got was a lot of stock very inexpensively and, of course, the risk of whether we would be able to make it both scientifically and economically," Faras said. They hired a dozen staff members, including four Ph.D.s, and by 1980 they were working on their first vaccines.

At the outset, they realized that $1.2 million was not enough to become "an effective player" in agricultural biotechnology, Faras said. To increase their revenues they entered into an agreement with American Cyanamid Co. Under a contract and licensing arrangement, MGI would develop agricultural products for American Cyanamid, which did not have an in-house biotechnology effort. American Cyanamid bought an equity position in MGI for $5.5 million and provided more than $3 million for research contracts that would range over four years. With these funds MGI was able to build a new laboratory facility and expand the staff to about 70, including over two dozen Ph.D.s and veterinarians.

### The Scientific Staff

MGI initially looked for scientists who could develop veterinary vaccines and other biologicals. They needed a team of scientists with expertise in molecular biology, immunology, and the chemistry of proteins and nucleic acids. Product development would require isolating the appropriate viral or bacterial genes and inserting them into a bacterial host that would then produce the vaccine protein.

To produce a commercial product, vast amounts of a vaccine protein are necessary, which means that genetically modified bacteria must be grown on a large scale in fermenters. Inside the fermenter, the bacteria express the foreign gene and produce the vaccine protein. Then that protein must be separated out of the mixture—at relatively low cost. For these tasks, MGI needed staff skilled in biochemical engineering and fermentation technology. Clinical testing of the genetically engineered vaccine is also needed.

When MGI began to move a couple of years ago into plant crop improvement, they also assembled a team of plant molecular biologists.

### *Financial Prospects*

At first, MGI was content to develop products for other companies, such as American Cyanamid, under a contract and licensing arrangement. "It was quite attractive; it gave us a nice net cash flow from contract research. The problem is that down the line you have to share the royalties, if you are a contractee, your share of the royalties is not always that good," Faras said.

"If you want to become a viable commercial entity, you have to think about producing and marketing those products yourself. To that end, we have started to hire marketing and sales distribution staff."

MGI's first product has recently passed clinical and field trials and is now being marketed in Canada. It is a monoclonal antibody for scours, or neonatal diarrhea, in calves. Other vaccines and monoclonal antibodies are now undergoing trials in the United States.

The scours antibody is bringing MGI "our first authentic sales revenues," Faras said. The company has also raised capital in other ways. In June 1982 it had its first public stock offering, followed by a second in March 1983. As separate ventures, MGI has also entered into a research and development limited partnership and a consortium arrangement with Martin Marietta, Inc.

From these activities, MGI has raised about $45 million since June 1982. This had helped to abate the "sea of red ink," as Faras jokingly described the financial situation attributed to new biotechnology firms. "With new revenues, the amount of losses we are incurring is looking better each month."

Overall, the biotechnology boom seems to be reaching equilibrium of sorts. Some companies, like MGI, are marketing their first products. Others have already failed. Given the numerous scientific uncertainties, some of the "survivors" may still drop out of the field.

# Safety
# Regulations

The debate about the safety of recombinant DNA research began almost as soon as the first such experiments were reported. In 1971 a molecular biologist proposed to combine DNA from a monkey tumor virus, known as SV40, with a plasmid from the bacteria *Escherichia coli*. This immediately raised fears among some scientists that the modified *E. coli*, containing monkey virus DNA, might somehow infect humans and cause cancer. Though this possibility seemed unlikely, it could not be dismissed for several reasons. First, *E. coli* commonly reside in the human intestine. If the recombinant molecule were inadvertently ingested by a human, it might be able to establish itself in the intestine. And second, though the virus has not been shown to cause cancer in humans, it does cause cancer in mice and hamsters and also causes human cells in culture to grow abnormally. After his colleagues voiced such concerns, the molecular biologist voluntarily deferred his experiment.

When a small group of molecular biologists met at a Gordon Conference in 1973, they again discussed the potential hazards of recombinant DNA experiments. After the meeting, they wrote a letter to *Science* to alert the broader scientific community to their concerns. In the letter they suggested that the National Academy of Sciences investigate the hazard, which it did. In 1974 a National Academy of Sciences committee recommended a worldwide moratorium on certain types of recombinant DNA experiments—such as those that would introduce into bacteria viral genes or genes that confer antibiotic resistance—until the safety hazards could be assessed. They also called for an international conference on the issue and suggested that the National Institutes of Health establish an advisory committee to develop safety guidelines for future recombinant DNA research. All three suggestions were followed.

The conference was held the following year at the Asilomar Center in Pacific Grove, California. By that time the scientists were not only concerned about the deliberate transfer into *E. coli* of a harmful gene, such as a gene from a cancer virus or a toxin, but they also wondered about the unforeseen hazards of combining genes of two different species—even if those genes were thought to be harmless. Since such recombinant organisms did not exist in nature, the scientists could not predict with accuracy what risk they might pose, not only to human health, but also to plants, animals, and the environment.

Nor was the debate confined to the scientific community. The public became increasingly concerned about both the safety questions and the moral and philosophical implications of the new technology. In creating novel organisms, scientists would have the power to alter the course of evolution. Many individuals, including some scientists, questioned whether scientists should be entrusted with such power. They also asked who should decide these issues—the scientists or the public.

Yet at Asilomar, the discussion was focused on scientific issues. The participants agreed that the moratorium should be lifted for the vast majority of recombinant DNA experiments, provided that appropriate precautions were taken. The safety strategy they suggested was that recombinant microorganisms be contained and that the level of containment correspond to the level of estimated risk of each experiment. Containment would be achieved through two methods: *biological*, the use of enfeebled strains of bacteria that could not survive outside of the laboratory; and *physical*, the use of laboratory procedures and equipment to prevent inadvertent release.

A committee of the NIH, now known as the Recombinant DNA Advisory Committee, or RAC, translated those recommendations into guidelines. These guidelines, adopted in 1976, specify the physical and biological containment conditions under which recombinant DNA experiments can be performed.

The guidelines are binding only for federally funded research. To date, industry has voluntarily complied with the guidelines, following procedures suggested in the guidelines for obtaining project approval.

In research conducted since 1979, the alleged hazards have not materialized. As knowledge accumulated, the guidelines have gradually been relaxed—that is, the containment levels required for certain experiments have been lowered. Now most experiments can be performed at the lowest biological and physical containment level.

Nonetheless, some safety questions still remain. One is the risk posed by the intentional release of novel organisms into the environment. Researchers have engineered microorganisms that, in the laboratory, can degrade a dioxin or others that might be used to clean up oil spills. The major uncertainty is whether they will disrupt the balance of the ecosystem in which they are released.

The concern is not with genetic engineering per se. The introduction of any species to an ecosystem it does not normally inhabit can have unexpected results. There are many examples of organisms that have become serious pests after they were released into a new area. In 1869 the gypsy moth (*Porthetria dispar*) was introduced into Massachusetts as part of a silk production experiment. It is now a serious forest pest in much of the Northeast. Prickley pear cactus (*Opuntia*) was introduced into Australia from Latin America and posed a serious threat when it spread into grazing land. The South American cactus moth (*Cactoblastis cactorum*) was deliberately introduced into Australia to bring the cactus under control.

## The Need for Continued Diligence

Ray Thornton was an active participant in much of the early debate over recombinant DNA regulation—both as a former U.S. congressman and as chairman of the RAC from 1980 to 1982. He is now president of Arkansas State University. As he described, he addressed another convocation at the National Academy of Sciences seven years ago.

"As I stand here, I can't help but have a sense of déjà vu. This room was filled with people concerned about whether there should be a moratorium on all recombinant DNA research. There were placards, there were protesters, there was heckling from the audience. It took a good bit of courage at that time to stand in this auditorium and suggest that the needs of science called for us to move forward cautiously and carefully in this area. That has changed over the past seven years, and today we recognize the enormous potential benefits that this new technology, this new way of doing things has made possible.

"Perhaps it may also require a bit of courage to suggest to this group today that there is still a continuing need to be aware of the safety, ethical, and moral issues of genetic engineering."

Thornton reminded the audience that these technologies provide an opportunity to direct the course of evolution, particularly to speed it. On a practical level, such changes can disrupt an ecosystem. On a philosophical level, this new ability may undermine man's reverence for life.

Many researchers and observers of the field think that if scientific work proceeds intelligently and prudently, and if scientific directions and developments are open to public scrutiny, then the safety issues can be resolved. Others, Thornton said, are less convinced that humanity is prepared to cope with the moral and ethical aspects of genetic engineering.

To date, scientists have been able to pursue recombinant DNA research with remarkable freedom, Thornton said. The use of voluntary guidelines—rather than legislation—is a novel approach and is far more flexible than the regulations governing the atomic, pharmaceutical, and chemical industries. He sees this freedom given to genetic engineering as a reflection of the public's confidence in the scientific community; of the public's belief that safety issues will be openly and honestly discussed.

"Only a handful of serious safety questions remain for RAC to consider," Thornton said. Among those is the release of genetically engineered organisms into the environment. "We're not talking about working with new organisms in the laboratory. We're talking about what recombinant life forms can be put in an oil well."

Other issues may emerge as the genetic engineering of plants nears application. In deciding what, if any, regulatory approach to take, the RAC or any other oversight body will need to draw on the knowledge of agricultural scientists, ecologists, and others. "One of the things that may have gone wrong six or seven years ago, that may have contributed to the public outcry over recombinant DNA research, is that the molecular biologists who were involved did not have the benefit of input from immunologists, epidemiologists, and others who could have helped them to assess the dangers. Because of this lack of knowledge, the restrictions initially applied were perhaps too severe. We have an opportunity to learn from that mistake. By drawing on the expertise of a number of disciplines, we can develop an approach that both satisfies the concerns for safety, yet does not unduly restrict the application of new research methods."

# Patents

As industry moved from genetic engineering research to product development, they encountered uncertainty about patent protection for their inventions. Clearly, the processes used in creating a specific recombinant microorganism were patentable. But what about the genetically engineered microorganisms themselves, such as those designed to degrade oil spills or combat soil-borne diseases? Could these novel life forms be patented?

For many years, Section 101 of the Patent Law has provided protection for any process, machine, manufacture, or composition of matter that meets certain criteria. Yet as Rene D. Tegtmeyer, assistant commissioner for patents of the U.S. Patent and Trademark Office, explained, it had been widely thought that the section did not apply to plants and other living things. In 1930 the Plant Patent Act was passed to allow patent protection for certain asexually propagated plants. The distinction was made between asexually and sexually propagated plants, because at that time it was not thought possible to produce a stable, uniform line through sexual reproduction. Those ideas were revised, and in 1970 Congress passed the Plant Variety Protection Act, which allowed protection of some sexually reproduced plants. The act specifically excluded bacteria, fungi, tuber-propagated plants, uncultivated plants, and first-generation hybrids.

In some situations, a genetic engineering company might prefer to rely on trade secret protection rather than apply for a patent. The advantage is that the trade secret need not be revealed and that such protection lasts indefinitely. The drawback is that if the secret is made public, either through legal or illegal means, then protection is lost. For these reasons, trade secret protection might prove inadequate or inappropriate for genetic engineering. Many of the products could be "reverse engineered"—that is, a competitor could use the product to deduce the original process. In addition, academic researchers might balk at trade secret protection, as they would not be able to publish their results.

Patents, too, have their drawbacks. Under patent law the applicant must have an invention that is new, useful, and unobvious and must describe the invention in sufficient detail to allow a person skilled in the field to use or operate it without undue experimentation. The disclosure requirement was a balance struck by Congress to encourage invention—both by protecting the rights of the inventor and by making new technical information publicly available so that others can learn from it. Patents provide the holder with exclusive rights to an invention for 17 years. During this time, if the patent is infringed, the holder has legal recourse.

In most cases, patents would seem to be the preferred method of protecting genetic engineering technology, Tegtmeyer said. "Inventions in this field have a good prospect for relatively long commercial life or usefulness though requirements for pre-marketing approval may be a factor in some areas. Competition will be heated; new entrants to the field will be many; market potentials are huge; and research and development is expensive. Genetic engineering is a suit tailor-made for the patent system." But first, the uncertainties had to be resolved.

## The Test Case

The test case was a patent application filed in 1972 by Ananda Chakrabarty, a General Electric scientist. By transferring plasmids from several bacteria into one bacterium, he had endowed this bacterium with the ability to degrade oil. His patent application included claims for the process used to engineer the new bacterial strain, for the carrier material to be used with the modified bacteria, and the genetically engineered bacterium itself.

The U.S. Patent and Trademark Office granted a patent on the process and the carrier but denied a patent on the bacterium on the grounds that living organisms are not patentable subject matter under Section 101. The Court of Custom and Patent Appeal then ruled that a patent could not be denied solely because the invention was a living organism. In its 1980 *Diamond* v. *Chakrabarty* ruling, the Supreme Court upheld the decision, affirming that genetically engineered microorganisms are patentable. The court ruled that "the relevant distinction was not between living and inanimate things but between products of nature, whether living or not, and human-made inventions." In other words, human intervention determines whether an organism is patentable subject matter.

Questions seem certain to arise concerning how much human intervention is necessary before an organism qualifies as subject matter that can be protected by patent. The Patent and Trademark Office has said that it will decide which genetically engineered organisms are patentable on a case-by-case basis.

Other issues may complicate the patenting of biotechnology products. For one, the applicant must demonstrate that the invention is novel and not obvious to someone having ordinary skills in the field. Yet "the very scope and complexity of the field and its rapid growth creates difficulty in determining what ordinary skill is at any time and what is obvious or novel," Tegtmeyer said. Similarly, the requirements for full disclosure of the "best mode" for using or creating the invention may also prove difficult. It may not be possible to adequately describe a microorganism or plasmid, or the genetic engineering processes. In those cases, a sample must be available to the public.

## Patent Activity

Difficulties aside, the Chakrabarty decision has greatly increased patent activity in the field of genetic engineering. A number of patent applications were suspended pending resolution of the case; those and new applications have since been processed. Now some 500 applications related to genetic engineering are pending. More are expected: patent

Top Assignees in Genetic Engineering (Class 435/172), 1979–1982

| Assignees | Percent |
| --- | --- |
| Upjohn Co. | 9.7 |
| Ortho Pharmaceutical Corp. | 7.1 |
| Ajinomoto Co., Inc. | 5.3 |
| President and Fellows of Harvard College | 4.4 |
| Regents of University of California | 4.4 |
| Research Corp. | 3.6 |
| Cetus Corp. | 2.7 |
| Genentech, Inc. | 2.7 |
| Noda Institute for Scientific Research | 2.7 |
| Agroferm Ag | 1.8 |
| All Other Organizations (44) | 47.0 |

SOURCE: U.S. Patent and Trademark Office.

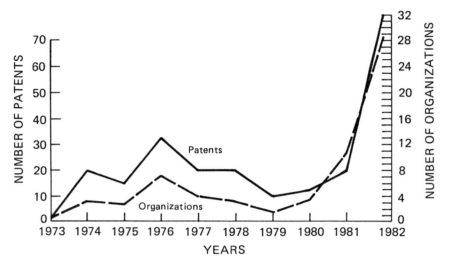

U.S. patenting activity on genetic engineering (Class 435/172): 1973-1982. Courtesy of U.S. Patent and Trademark Office.

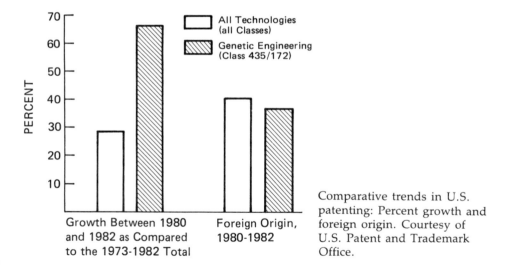

Comparative trends in U.S. patenting: Percent growth and foreign origin. Courtesy of U.S. Patent and Trademark Office.

activity in genetic engineering is increasing at roughly twice the rate of other technologies.

A sizeable portion of U.S. patents are issued for inventions developed in other countries. From 1980 to 1982, the percentage of foreign patents issued for genetic engineering was lower than that for other technologies. "This seems to indicate a concentration of genetic engineering inventiveness in the United States at this time," Tegtmeyer said.

# Index